MOLECULAR
BIOLOGY
INTELLIGENCE
UNIT

THE MEANING OF NUCLEOCYTOPLASMIC TRANSPORT

Paul S. Agutter, Ph.D.

Napier University
Edinburgh, Scotland

Philip L. Taylor, Ph.D.

Edinburgh, Scotland

Springer-Verlag Berlin Heidelberg GmbH

R.G. LANDES COMPANY
AUSTIN

MOLECULAR BIOLOGY INTELLIGENCE UNIT

THE MEANING OF NUCLEOCYTOPLASMIC TRANSPORT

R.G. LANDES COMPANY
Austin, Texas, U.S.A.

International Copyright © 1996 Springer-Verlag Berlin Heidelberg
Originally published by Springer-Verlag New York Berlin Heidelberg London Paris
Tokyo Hong Kong Barcelona Budapest in 1996
Softcover reprint of the hardcover 1st edition 1996

 Springer

International ISBN 978-3-662-22504-2

While the authors, editors and publisher believe that drug selection and dosage and the specifications and usage of equipment and devices, as set forth in this book, are in accord with current recommendations and practice at the time of publication, they make no warranty, expressed or implied, with respect to material described in this book. In view of the ongoing research, equipment development, changes in governmental regulations and the rapid accumulation of information relating to the biomedical sciences, the reader is urged to carefully review and evaluate the information provided herein.

Library of Congress Cataloging-in-Publication Data

Agutter, Paul S., 1946-
    The meaning of nucleocytoplasmic transport/ Paul S. Agutter, Philip L. Taylor
        p. cm. — (Medical intelligence unit)
    Includes bibliographical references and index.
    ISBN 978-3-662-22504-2   ISBN 978-3-662-22502-8 (eBook)
    DOI 10.1007/978-3-662-22502-8

    1. Nuclear membranes. 2. Biological transport. 3. Messenger RNA--Physiological transport. 4. Proteins--Physiological transport.
1. Taylor, Philip L. 2. Title. 3. Series.
QH601.2.A38 1996
574.87'32--dc20   96-22042
    CIP

PUBLISHER'S NOTE

R.G. Landes Company publishes six book series: *Medical Intelligence Unit, Molecular Biology Intelligence Unit, Neuroscience Intelligence Unit, Tissue Engineering Intelligence Unit, Biotechnology Intelligence Unit* and *Environmental Intelligence Unit.* The authors of our books are acknowledged leaders in their fields and the topics are unique. Almost without exception, no other similar books exist on these topics.

Our goal is to publish books in important and rapidly changing areas of bioscience and environment for sophisticated researchers and clinicians. To achieve this goal, we have accelerated our publishing program to conform to the fast pace in which information grows in bioscience. Most of our books are published within 90 to 120 days of receipt of the manuscript. We would like to thank our readers for their continuing interest and welcome any comments or suggestions they may have for future books.

Deborah Muir Molsberry
Publications Director
R.G. Landes Company

CONTENTS

A book's title should encapsulate its aims and scope, and the preface should make those aims and that scope explicit. So what is our title supposed to encapsulate? 'Meaning' is a notoriously difficult word, and as we shall try to explain here, we intend it in two different though related senses. 'Nucleocytoplasmic' is more straightforward, and for the time being we can take it as unambiguous. But 'transport', which is often assumed to be nonproblematic, is fraught with difficulty. That difficulty, the question of how we should properly interpret and use the word 'transport', might be considered the central theme of the book. In part, therefore, the title can be paraphrased as "What the word 'transport' denotes, or should be taken to denote, in the context of 'nucleocytoplasmic transport'". However, we also intended to convey something like "The relevance of nucleocytoplasmic transport research to other, sometimes apparently unrelated, aspects of eukaryotic cell biology." The connection between these two senses of 'meaning' is not immediately obvious, so we need to explain it.

We can begin with a few uncontentious claims. First, based on publication and citation evidence, nucleocytoplasmic transport is an active and much-reviewed field of research in modern cell biology. Insofar as a book is just one more review among many, it behooves the authors to take a novel approach, to try to present the evidence from a new and (with luck) helpful perspective. Second, books are generally intended for wider audiences than review articles, so an attempt to interest nonspecialists in the research field seems appropriate, and a serious effort to relate the field to other cell-biological interests seems a reasonable means to that end. Third, those discoveries about nucleocytoplasmic transport that are now becoming textbook knowledge are of at least some general interest, so our ambition is not necessarily unrealizable. For example, if the focus of interest is a certain protein that spends most of its life in the cytoplasm but under certain conditions enters the nucleus, most cell biologists today will readily accept that a nuclear location signal is somehow involved. Several hypotheses might be proposed. The protein may contain a cryptic location signal which is exposed by a conformation change under the new cellular conditions; or it might contain a noncryptic location signal but remain bound to a cytoskeletal element or cytoplasmic membrane until conditions change; or again it might contain no location signal at all, but form a stable heterodimer with a newly synthesized protein that does have such a signal. In pursuing such hypotheses, a researcher is likely to gain insights into the molecular and cell-regulatory mechanisms of interest. New knowledge about nucleocytoplasmic transport can also attract wider attention through the sense it makes of otherwise incompletely understood aspects of cell behavior. For example, the growing belief that mRNA export from the nucleus depends on the integrity of the 5' methyl cap seems to add a clear functional significance to what we know about the capping process. On a different level, nucleocytoplasmic transport is interesting because its machinery must have been present in the earliest eukaryotes. This machinery (e.g., sequences of pore-complex components) has shown a high level of evolutionary conservation. We can therefore expect further understanding of nucleocytoplasmic transport to throw some light on the origins of eukaryotes.

These reasons for interest in a specialized research field are (though not negligible) rather limited, and might not be sufficient to justify, say, the time spent in reading a book devoted to that field. After all, if there is no more in principle to nucleocytoplasmic transport than the (indirect) interaction between an intramolecular signal in the transported molecule and a component of the pore complex, followed by engagement of the machinery that translocates the molecule of interest across the nuclear envelope, then the scope for overlap with other aspects of cell biology is inherently limited. As we see it, the catch in this argument is the 'if'. Suppose there is more to nucleocytoplasmic transport than events at the pore complex? Suppose transport involves structures and events quite remote from the nuclear envelope; structures and events that are investigated by researchers in fields usually considered irrelevant to transactions at the nuclear periphery? Then the conclusion might be very different, and the scope for overlap much wider. This is why, in our view, the range of potential interest in nucleocytoplasmic transport is a function of what 'transport' actually denotes, and this is the connection between our two senses of 'meaning'. We are persuaded that there really is more to transport than events at the pore complex, and the potential range of interest is correspondingly wide.

We intend to tackle this issue directly. For this reason, our first chapter might look more like an essay in semantics or linguistic philosophy than a survey of an area of cell biology. Its theme is the various uses of the word 'transport' in cell biology. By exploring these usages we expose a number of almost universally accepted but challengeable presumptions in our thinking, and these suggest connections between nucleocytoplasmic transport research and various other aspects of the discipline. These connections, and the exposed presumptions, dictate the organization and content of the remainder of the book. Either we shall convince the reader that 'transport' needs to be reconceptualized and that nucleocytoplasmic transport is ipso facto a field of general interest and significance, or we shall not. If the former, we shall be gratified; if the latter, we shall take comfort in the belief that we have either provoked a scientific debate, which is usually constructive in the long run, or been rebutted; and the refutation of error is reputedly the surest route to scientific progress. Whether readers accept our arguments or not, we hope, at least, that they will find our text reasonably informative and enjoyable to read.

CHAPTER 1

PERSPECTIVES IN TRANSPORT

'TRANSPORT' IS USED IN SEVERAL DIFFERENT WAYS IN CELL BIOLOGY

A word taken from everyday language and given a role in scientific discourse is often open to errors or difficulties of interpretation. Students learning about Newtonian mechanics for the first time find it difficult to cope with the specialized, restricted use of the word 'force', which clashes somewhat with their everyday use, especially if they have seen Star Wars. And no doubt the perceptions of budding particle physicists are clouded by the everyday connotations of 'color' and 'charm', not to mention 'strangeness'. As biologists, we are apt to forget that a word such as 'transport' also carries a host of connotations from everyday language. We sincerely believe that we have given the word a precise, connotation-free significance; we have not made it quantitative perhaps, but have otherwise taken a generalized everyday language definition and consistently applied it in our scientific discourse. We have performed much the same service for it as Newton performed for the word 'force'. A few moments' reflection might make us less certain of our semantic purity, and question exactly what phenomena a particular instance of 'transport' denotes.

In everyday language, 'transport' signifies the active movement of goods or persons from point A to point B. There is a definite starting point and a definite destination, and the process is directed and requires energy. Sometimes transport requires crossing a boundary, such as a national or state border, but that is contingent and forms no part of the definition of the word. Whatever or whoever is being transported is loaded onto the conveyance (e.g., aircraft, mule, shoulder, elevator) at point A and unloaded again at point B. The use of the word is very varied, but it always conforms to this generalized definition, so this, presumably, is the definition we have imported into cell biology.

But when we describe the uptake of glucose by a eukaryotic cell as 'glucose transport', what do we really mean? In a certain limited sense, it can be claimed that 'transport' here does conform to its generalized, everyday language definition. The plasma membrane, we say, contains a glucose transporter (the conveyance), which (in at least some of the best-studied cases) is energized by coupling to transmembrane ion gradients. Point A is a point on the outer surface of the plasma membrane contiguous to the transporter, and point B the corresponding location on the inner surface. This technical use of 'transport' (see virtually any modern textbook, for example ref. 1) therefore appears superficially consistent with its general everyday language definition; starting point, destination, conveyance and energy utilization are all present.

This claim of conformity, however, is a mere artifice, and does not withstand scrutiny. First, even if there were no energy utilization, the vast majority of biologists would still consider the word 'transport' to be legitimate and appropriate. Indeed there are cases of so-called 'facilitated transport', in which the conveyance is not energized. (The bizarre oxymoron 'facilitated diffusion' is sometimes applied in these cases. If 'diffusion' is intended scientifically here, this phrase has about the same semantic content as 'pantomime crankshaft' or 'thermonuclear napkin'. If it is not intended scientifically, why is it used in scientific discourse?) However, this is a minor point; second, and more important, we do not always (or even commonly) use the phrase 'glucose transport' to denote simply the step between molecule-on-outside-of-membrane and molecule-on-inside-of-membrane. Far more often, we intend to convey a message about glucose moving en masse from a state of solution in the extracellular medium to a state of solution in the intracellular medium, as indicated by the inclusion of glucose concentration gradients in our discussions. Now 'extracellular/intracellular media' are not exact locations; they do not have the characteristics of precise starting points and destinations. Indeed, we regard glucose (or any other molecules) in solution as having *no* exact locations; their distributions, we say, are random, by virtue of kinetic theory considerations.

Third, following from this, it is implicit in our usage of the phrase, and our attempt to account for glucose transport exclusively in terms of a membrane-crossing event, that the movement of glucose molecules in the aqueous phase, and ipso facto their arrival at and departure from the transporter, is orders of magnitude faster than the membrane-crossing event itself. And fourth, if the membrane were removed, but the concentration difference between intracellular and extracellular compartments were maintained, and the kinetics of exchange between these compartments were also unaltered, the continuing use of the word 'transport' would be deemed controversial at best. (Of a 'random' sample of twenty biologists presented with this scenario, three said they would retain the use of the phrase 'glucose transport', twelve said they would not, and five remained undecided even after a few drinks; P.S.A., unpublished observations.) Contrary to first appearances, therefore, 'transport' in the phrase 'glucose transport' has taken on a denotation quite at variance with its everyday language use. It does not entail a definite starting point or destination, it *does* entail the existence of a barrier to be crossed, and there is no necessary coupling to an energy supply; and, importantly, the 'real' transport event is taken to be the slow, specific, conveyance-dependent step, while the fast, random, and conveyance-independent steps are disregarded, precisely because they *are* fast, random, and conveyance independent. There is nothing at all wrong with using 'transport' in this specialized way, so long as we are aware of its divergence from everyday language use and can warn our students about it; but consistency of definition and denotation is usually considered a good thing, and in cell biological uses of 'transport', we do not, alas, find that consistency.

'Axonal transport' is a commonly-used phrase, which in itself conveys a consistent image to a large consensus of cell biologists; but the very structure of the phrase should warn us that it denotes something very different in kind from 'glucose transport'. It contains an adjective that defines a *direction* or *route*, and contains no reference to the substrate of transport, as per 'glucose' in 'glucose transport'. Here, we find ourselves genuinely close to everyday language use. There is a definite starting point (the neural cell body), there is a destination (the axon terminus), a conveyance (the transmitter vesicle interacting with the axonal microtubules), and the process is energized by a specific motor.[2] True, the exact locations of transmitter molecules are uncertain; they are somewhere inside the vesicle, presumed to be in solution; but 'axonal transport' refers to the vesicle itself, not individual transmitter molecules. Moreover, there is no sense of boundary-crossing implicit in 'axonal transport'; no membrane or other obvious barrier to vesicle migration intrudes between the cell body and axon terminus. It is quite evident that the word 'transport' appears as two distinct homonyms in 'axonal transport' and 'glucose transport', much as the word 'board' does in 'notice board' and 'parole board'. If we wish to explain what an 'emery board' is, we cannot rely on either 'notice board' or 'parole board' to provide an apt metaphor; and if we wish to know what 'nucleocytoplasmic transport' is, we cannot necessarily trust either 'glucose transport' or 'axonal transport' to provide an image that will not simply mislead us.

Nor should we infer that we have a choice between just two alternative denotations of 'transport' in cell biology. There has been considerable debate in the literature about the mechanisms whereby plasma membrane and secretory proteins are conveyed from the endoplasmic reticulum to the cell periphery. The controversy surrounding the bulk-flow hypothesis[3] has been treated as largely methodological.[4,5] We have no reason to doubt this, and no wish to enter the debate, but it is worth noting that 'transport' in this context (and the word is certainly used) has a denotation intermediate between those of the two homonyms we have discussed so far. For instance, there is a fairly specific starting point (the translation site and point of uptake into the ER), but not necessarily a precise destination; in terms of the individual cell, the destination of a secreted protein is undefined. There is a conveyance system, comprising the endomembranes of ER and Golgi. However, the bulk-flow hypothesis (and at least some alternatives) do not allow the process to be specifically energized. Whether the formation of cis-Golgi vesicles and further transmission through the Golgi can be construed as barrier crossing is not clear,[6] but the bulk-flow hypothesis makes no such implication. This example serves to suggest that 'transport' occurs in cell biology not as a set of two or more sharply distinct homonyms, but rather as a denotational continuum having several semantic dimensions, with one pole of the continuum (the 'axonal transport' pole) coinciding closely with the generalized everyday language definition. The point, or region, of this continuum to which 'nucleocytoplasmic transport' belongs is not evident a priori, so to make presumptions about the exact denotation of the phrase would be foolhardy.

Because of our focus on details of actual word use, this preamble, and much of what follows, leaves us open to the charge of fudging our own arguments by skillfully deploying several notoriously variable words; specifically, 'hypothesis', 'model', 'perspective', 'methodology' and 'paradigm'. An exhaustive discussion of each of these words would scarcely aid our progress and would surely irritate the reader, but brief indicators of our intended uses would be helpful. A 'hypothesis' is a statement intended to explain a set of data, testable (i.e., open to refutation or corroboration) by observation or experiment. A 'model' is an attempt to represent the system under study, stripped of all features that are complicating and regarded as superfluous or irrelevant; it provides the basis or context for hypothesis formulation. A 'perspective' is a way of thinking about or visualizing the phenomena of interest that allows us to decide, provisionally, which features of the system are important and which are not, and thereby to articulate one or more models. A 'methodology' is an approach to experimental or observational study; it is linked with perspective, in that it comprises a set of principles whereby methods can be devised that will elucidate the 'essential' features of the system under study, but not contaminate the picture with data about 'superfluous' ones. We use 'paradigm' more or less in its Kuhnian sense, as the sum of perspective and methodology.

'NUCLEOCYTOPLASMIC TRANSPORT' AND 'GLUCOSE TRANSPORT' HAVE SOME COMMON CHARACTERISTICS, BUT SOME MAJOR DIFFERENCES

Whatever controversies might abound in the nucleocytoplasmic transport field, there are some beliefs from which there is no serious dissent. Material that moves from one major compartment to the other in a eukaryotic cell crosses a barrier, the nuclear envelope, which comprises two membranes and a lamina, within which pore-complexes are set.[7] The outer (cytoplasm-facing) membrane resembles rough endoplasmic reticulum. The lamina bears broadly the same relationship to the inner membrane as the spectrin-actin framework does to the erythrocyte membrane. The pore complexes are large (approximately 100 nm diameter) cylindrical structures which pass through both membranes and the lamina and extend into the nucleoplasm and cytoplasm. We shall discuss some structural details in chapter 3, but this outline description is enough to establish that the nuclear envelope is an obvious physical barrier between the compartments; and the appearance of the pore-complexes suggests that they are the main sites of exchange between cytoplasm and nucleoplasm, a hypothesis that has been amply corroborated for more than three decades.[8,9] Indeed, it has been firmly established that the pore-complexes allow free passage of low molecular mass solutes, and nucleocytoplasmic transport research therefore focuses exclusively on macromolecules and complexes: proteins, RNA and RNPs.[10,11] These macromolecules have been shown to pass through the pore complexes by mechanisms that involve specific recognition (part of the transport substrate, i.e., a portion of the amino acid or nucleotide sequence acts as a signal, which the pore-complex indirectly recognizes); and, in at least some cases, energy transduction (for example refs. 12-15).

Again, we shall discuss details later; our present business is with generalities. Macromolecular movement between nucleus and cytoplasm is in some respects like glucose uptake by cells. There is a physical barrier; there is a specific conveyance system for enabling the transport substrate to cross that barrier; and in at least some cases, the conveyance system is energized. For many researchers in the field, these resemblances have sufficed; nucleocytoplasmic transport processes are conceptualized by analogy with glucose transport. That is, nucleocytoplasmic transport is presumed to consist of *nothing but* the interaction of the transport substrate and its effectors with the pore-complex, and the subsequent (and perhaps consequent) translocation event.[12,15]

When we recall the implications of the 'glucose transport' model, it is easy to see that this conceptualization is inadequate. We are able to think of glucose transport exclusively in terms of membrane crossing only because: (a) glucose molecules in solution have no exact locations and (b) glucose molecules move much more rapidly in the aqueous phase than they do through the membrane. It is these two assumptions that justify us in speaking as if an individual molecule was transported between a starting-point and destination which coincide with the extreme positions of the transporter at the membrane surfaces; or rather, as if the isolated membrane-crossing event and the bulk movement of glucose between aqueous phases were to all intents and purposes equivalent. However, no such assumptions can be made about macromolecule transport inside cells. For instance, a gene transcript might remain in the nucleus for several minutes, during which time it is processed to mature RNA, but the mRNA is translocated through the pore-complex in seconds at most.[16] It is not the case that passage through the barrier is much the slowest step in overall migration. Also, messengers and their precursors are not spread diffusely throughout the nucleoplasm; however we interpret the data, these molecules have fairly definite locations.[17,18] Similar observations apply to other macromolecules and complexes that pass between nucleus and cytoplasm. It follows that the implicit postulate of the glucose transport analogy, that bulk transport can effectively be equated with transport of individual molecules and can be construed as if it began at one end of the pore complex and ended at the other, is completely unfounded. It is seriously inconsistent with the evidence. Whatever nucleocytoplasmic transport is, it does not resemble glucose transport.

This inference, which depends only on a logical evaluation of commonly-agreed evidence, has a number of startling implications. First, it means that we must be prepared to think of nucleocytoplasmic transport as including events within both the nucleoplasmic and cytoplasmic compartments; not as parenthetic supplements to events at the pore-complex, but (potentially) as of equal significance to translocation, if not greater significance. Second, since these events coincide with other activities, their relationships with these activities become material considerations. For example, while mRNA and its precursors are moving within the nucleus, a movement which we have shown must be considered *part of*, not a prelude to, mRNA transport, they are being processed. It follows that posttranscriptional processing and the early stages of mRNA transport are simultaneous and presumably intimately associated events. This makes the studies of mRNA transport and posttranscriptional processing logically and biologically inseparable, immediately extending the range of possible interest in nucleocytoplasmic transport research. Third, it is at least a reasonable guess that long-range structures in the nucleus and cytoplasm are implicated in transport; nonrandom distribution and slow migration could sensibly be interpreted in terms of association of the transport substrates with cytoskeleton or nucleoskeleton, which would immediately link research on these structural entities with the nucleocytoplasmic transport field. Indeed, associations between cytoplasmic messengers and the cytoskeleton are now well established.[19,20] Fourth, it raises a wider question about the nature of macromolecule movement within cells. It is commonly assumed that proteins, for example, diffuse through the solution phase within cells, unless they are associated with solid structures such as membranes or cytoskeletal elements; much as glucose is held to diffuse in solution. If this assumption is incorrect when applied to nucleocytoplasmic transport, then it is presumably generally incorrect. This constitutes a challenge to diffusion theory itself, at least as applied to macromolecules in cells, and this has fundamental implications for cell physiology that need to be explored. The restricted mobilities of cytoplasmic proteins, nuclear proteins and RNAs have been well demonstrated.[21-23] Fifth, the following question cannot be avoided. The 'glucose' analogy for nucleocytoplasmic transport is easily refuted, as we have seen, yet it is tacitly accepted by the majority of workers in the field, and questioning of it has not been taken seriously by many. Why?

THE GLUCOSE TRANSPORT ANALOGY PERSISTS, BUT THE REASONS FOR ITS PERSISTENCE ARE LARGELY UNSOUND

This a difficult question to address, because it is not one that researchers willingly debate, at least in print; it is not generally considered pertinent to new developments and discoveries, so it is neglected. The discussion that follows is based on informal conversations. The participants would probably wish to remain anonymous, but we have endeavored not to misrepresent them.

First, there is the Ockham's Razor argument. There is no doubt about the involvement of the pore-complex in transport, and no doubt about the significance of intramolecular signals in the transport substrates. These are therefore proper foci of research. Nothing else in the cell can be implicated so unequivocally. The principle of simplicity therefore guides us to retain the narrow focus and to neglect everything but pore-complexes and signals. On the face of it, this appears to be sound scientific practice, but it should be remembered that Ockham's Razor is often a two-edged weapon, and has on occasions been deployed as a dangerous blunt instrument. It would be doubtful practice to construe a transcriptional event as involving only the gene, an upstream TATA box and RNA polymerase II, on the grounds that these were indubitably involved, while the constellation of possible promoters and enhancers was too ill-defined and open-ended to merit consideration 'at least at this stage'. A piece of research predicated on this attitude would soon become unproductive, and perhaps seriously misdirected, and we would be justified in saying that the researcher had used Ockham's Razor to cut her own throat, if not to bludgeon herself on the head.

Second, there is the supplementary argument that the broader perspective which we argue is appropriate for nucleocytoplasmic transport studies invokes ill-characterized structures about which no clear assertions can be made. Since one of these structures appears to be the actin cytoskeleton,[20] we suspect that there are cell biologists who would cavil at the description 'ill-characterized' and would wish to make several clear assertions. However, let us suppose that the structure referred to in the argument is the nuclear matrix. This raises a complicated issue, which we shall address later (chapter 3), because we believe there are good grounds for implicating the nucleoskeleton in transport;[24,25] so at present we shall confine ourselves to general replies. The adjective 'ill-characterized' is deliberately if mildly pejorative, so it has an element of value judgment as well as description. As far as its descriptive character is concerned, what exactly does it indicate? Like 'transport', it is a term that requires analysis; it will no doubt turn out to denote different things in different contexts, but it is important to know just what it does signify in any instance, otherwise discussion will lack usefulness. We might add the rather cheap but not unjustified barb that the pore-complex is also ill-characterized, an assertion that must be subjected to clarification it its turn.

Third, there is an argument by analogy. Low molecular mass solutes transverse the pore-complex as readily as a fly passes through an open window. They show no obvious restrictions of mobility in either nucleoplasm or cytoplasm, or at any point in between. Therefore, larger molecules presumably behave in just the same way, except that (being bigger) they experience some difficulty in crossing the pore-complex and need specialized machinery to help them. The issue of transport therefore concerns only the events at the pore-complex. Perhaps we should emphasize that a fly differs from (say) an albatross in several respects apart from its linear dimensions. The analogy between small solutes and macromolecules was explored experimentally in the early days of research in the field, and found to be wanting;[26,27] indeed, these studies first indicated the existence of specific transport systems. Analogy-as-heuristic plays an important role in science, but analogy-as-justification has its proper home, if anywhere, in political rhetoric.

Fourth, work that has been accomplished with the tacit assumption that nucleocytoplasmic transport can be conceptualized in the same way as glucose transport has yielded a number of impressive, indeed exciting, results, which form virtually the whole of our current understanding in the field. This is undeniable. However, results are produced by the application of methods, not of perspectives or models; perspectives influence the choice of methods, but do not directly influence the data. Interpretations of results, on the other hand, are likely to be strongly influenced by models, and will in turn change the articulation of those models, perhaps eliminating some. A number of implications follow. During the course of this book we shall inevitably reinterpret some data, the authors of which have assumed a model that we believe to be logically unsound. We shall try to be explicit about these effects of perspective clash, but it is a moot point whether radical reinterpretations of data without authorization from the legitimate holders of the intellectual property rights thereto constitute 'fair reviewing'. As will become apparent, this consideration has not inhibited us. Likewise, we shall interpret certain controversies in the field in frankly tendentious ways, and we should make it clear now that our position does not imply any denigration of the (often very high) quality of research conducted by those whose interpretations we cannot accept. On the issue of methods, we face an immediate and important question: which methods are legitimized by which perspectives, and in what ways (if any) has the widespread adoption of the narrow 'glucose transport' perspective influenced the choice of methods, and hence the range of data available?

So far as we have been able to ascertain, these four are the only defenses of the 'glucose transport' perspective currently on offer. The first three are clearly unsound. The last is more interesting, and valid as far as it goes, but it raises issues of both interpretation and method that need to be pursued further.

NO ALTERNATIVE PERSPECTIVE IS JUSTIFIABLE A PRIORI

In case it should be inferred from the foregoing that we have a definite alternative perspective in mind, let us be quite clear that this is not the case. The main purpose of this book is to explore the field in the hope of finding an appropriate perspective, perhaps even a model, that will suffice. All that we have achieved so far in this chapter is to show that the most widely-accepted perspective is fatally flawed, and in doing this we have given some credence to our claim that the study of nucleocytoplasmic transport overlaps materially with many other fields of cell biology and is therefore potentially interesting to workers in those fields.

The supposition that we are beating a particular drum might gain force from the fact that one of the present authors outlined a 'solid-state transport' perspective in the field around 1980 (for example ref. 28). This view of nucleocytoplasmic transport, specifically relating to mRNA, has been featured in several reviews,[14,16] and mention of it is needed here. In brief, it holds that mRNA and its precursors remain attached to solid structures throughout their active lives in the cell, and are moved from place to place not in the solution phase, but by transfer along the framework, or by net movement of the framework itself. Whether or not this view can be justified in the long run, it is inadequate as it stands. First, it does not constitute a model, and gives rise to no directly-testable hypotheses. Second, although its case can be argued in some detail, much of the argument rests on certain interpretations of data that are not necessarily correct. Third, solid-state transport has routinely been envisaged as comprising three conceptually distinct stages: transfer from intranuclear binding sites to the pore-complex (*release*), *translocation* across the pore-complex, and *cytoskeletal binding*. For one thing, this division makes intranuclear and intracytoplasmic events in transport appear both homogeneous and passive, and they are almost certainly neither. For another, it fails to discriminate between binding to the pore-complex and movement through the pore-complex, and these events are experimentally distinguishable. Finally, the possibility that the perspective can be extended to cover other substrates of nucleocytoplasmic transport, such as tRNAs and proteins, has not been adequately explored.

One possible model representing this perspective would be an analogy with axonal transport, the other pole of our denotational continuum from glucose transport. But this is clearly flawed, because a barrier, the nuclear envelope, is involved (which is not the case in axonal transport); and even if there is a continuous nucleo-cytoplasmic fibrillar system, which is a contentious claim, it certainly does not consist of a single fiber type analogous to axonal microtubules, so it is scarcely possible to conceive of a homogeneous population of motor-fiber interactions. Moreover, it is far from clear that all nucleocytoplasmic transport processes are energy-requiring, or at least that they remain so through nucleoplasm, pore-complex and cytoplasm. Obviously the 'axonal transport' analogy is no more credible than the 'glucose transport' analogy. Some alternatives to the 'axonal transport' analogy have been proposed,[29] but they fare little better in the face of general criticisms, and it is doubtful whether all of them can be regarded as 'solid-state models' in the original sense of the term.

It is both necessary and, in this situation, inevitable, that we approach this survey of nucleocytoplasmic transport without fixed preconceptions. The main value of the solid-state perspective has been to question the tacit consensus viewpoint; that mission accomplished, it should be subjected to as much critical reevaluation as the 'glucose transport' analogy. Preconceptions prejudice the quest for meaning.

A METHODOLOGICAL PARADOX

Alternative methodologies, or approaches to experiment design, have a bearing on the choice of perspective, and the perspective plays a large part in selecting them. We might expect different perspectives to select different methodologies, and this appears to be borne out in practice. Proponents of one perspective are likely to reject the methodology of another, which again seems to be the case. However, the uninitiated might suppose that proponents of the reductionist, Ockham-whittled 'glucose transport' viewpoint would opt for studies on cell fractions, such as isolated nuclei and envelopes, while apologists for the holistic, cluttered 'axonal transport' position would demand studies on whole cells; and this is more or less the reverse of the real situation. An account of this seeming paradox will complement our discussion, and will serve to introduce general issues about experimentation without recourse to the otiose device of devoting a separate chapter to a detailed survey of methods. In this account, some of the difficulties we have already encountered will reappear in a new guise, and become more fully articulated in the process.

In the 1980s, techniques became available for genetically altering RNA and protein sequences at specific points, microinjecting
the wild type or mutated molecule into a single cultured cell, and
locating it at various subsequent times by, for instance, the use of
suitable monoclonal antibodies. It was found that altering certain
small parts of a sequence prevented (say) the transit of a mutant
nuclear protein from cytoplasm to nucleus, and conversely, that
the introduction of the wild type form of this oligopeptide into a
normally cytoplasmic protein caused it to adopt a nuclear location. This is how nuclear location signals were first identified,[30,31]
and for a time at least the search for location signals on nuclear
proteins became something of a boom industry.[13,31,32] This important achievement confirmed the value of molecular biological techniques in the study of nucleocytoplasmic transport, and showed
that by the use of such techniques location signals can be identified; and it proved possible to establish the relative importance of
each individual residue in the signal for nuclear uptake, and the
tolerance of the molecule to mutation at that residue.[30] The much-
overused word 'breakthrough' seems appropriate here, for certainly
the dramatic increase of research activity in the nucleocytoplasmic
transport field dates from around that time. The effect was to establish the genetic manipulation/microinjection/identification approach as the paradigm methodology and to make methodological
alternatives inherently questionable. It seemed obvious that the cell,
remaining intact after microinjection, is minimally perturbed, so
the physiological norm is maintained and the results can be deemed
substantially free of artifacts.

There is an assumption here that needs to be made explicit. In
a spherical cell with a diameter of 30 μm and a 10 μm diameter
nucleus, there are something like 8-10 pl of cytoplasmic water and
rather less than 0.5 pl of nuclear water. As we have already seen,
the endogenous macromolecules do not seem to be randomly dis-
persed in this water, but are mostly restricted in mobility for reasons that may have directly or indirectly to do with the all perva-
sive fibrillar and membrane structures of the cell. What volume of
liquid is introduced into the cell in a microinjection experiment?
If it is substantially less (say by two orders of magnitude) than
10 pl for a cytoplasmic microinjection, or 0.5 pl for a nuclear one,
then we may be justified in regarding the cell as minimally per-

turbed. Of course, a volume of unstructured liquid has been introduced, a demolition site in a well-planned townscape, but if it is small then normal traffic flow need not be seriously disrupted. However, if the microinjected volume is larger, then the potential for disturbing the cell's normal behavior is correspondingly greater. What is most relevant here is the uncritical supposition that a test molecule dissolved in this structurally vacant solution will immediately behave like its endogenous counterpart. Unless endogenous and exogenous macromolecules behave inside the cell like ordinary solutes obedient to the law of diffusion, and we have already suggested that they do not, then this supposition seems absurd. Conversely, those who accept the supposition are a fortiori committing themselves to the belief that macromolecules move around in nucleoplasm and cytoplasm by, to all intents and purposes, diffusion. (Parenthetically, we note that when the cell is subsequently permeabilized to allow ingress of a monoclonal to locate the test molecule, its integrity is most certainly challenged; but that is not the main point of our present argument.) The inference is clear: reliance on in situ methods as the only trustworthy and artifact-free sources of data about nucleocytoplasmic transport logically entails commitment to the 'glucose transport' perspective, which we already know to be flawed.

Lest it be thought that we are damning most of the best research in the field out of hand, let us be quite clear that no one doubts the validity of the data pertaining to identification and sequencing of location signals, or that these data could only have been obtained by the in situ approach. The issue does not have to do with the relevant features of the test molecule. It has to do with the supposition, which the foregoing discussion shows to be highly suspect, that a location or transport signal in the test molecule is recognized uniquely (though indirectly) by sites that are confined to the pore-complex, and that this is the only process directly relevant to migration between compartments. This supposition enters interpretations of the data automatically if the 'glucose transport' perspective is adopted, but not if it is not. One way in which we shall be obliged to reinterpret data is now clear; we cannot justify this supposition. Two other points can be made here. One, which is obvious, is that the adoption of in situ methods has contributed largely to the survival of the 'glucose transport' perspective, and this supporting act is reciprocal. The other, which is a little less obvious, is that in situ methods give reliable qualitative data about the test molecules, but are very difficult to make quantitative. The paradigm methodology has therefore told us little about the kinetics of nucleocytoplasmic transport, and (for much the same reason) has been rather uninformative about energy requirements and the relevant energy-transducing machinery.[33]

In vitro experiments using isolated nuclei or resealed nuclear envelope ghosts in aqueous suspension[34] could, in principle, make good these deficiencies and have been claimed to do so by certain protagonists, but they are generally disregarded. It is worth exploring just why they are disregarded, and how and under what conditions they might attain to a degree of credibility. In general, the in situ approach being now adopted as the methodological standard, it provides the criteria against which any in vitro method must be assessed. Any disparity in the results is attributed to in vitro artifacts and is taken as evidence that the method is unsatisfactory. To judge from referees' reports received by some of our colleagues around the world, authors of the in vitro studies are then exhorted to be 'more critical'. This irony is not intentional, so far as we can ascertain, but certainly the argument underpinning the exhortation does not itself withstand critical scrutiny.

The argument is this. When nuclei are isolated in bulk in aqueous media, their envelopes are torn and at least some pore-complexes are damaged.[35] Certainly many abundant proteins such as nucleoplasmin are lost.[23] The isolated nucleus is like a ruined house, lacking doors, windows and parts of its roof, structurally unsound, and retaining only its floors and walls, together with entities such as wallpaper that remain firmly attached thereto. Such an object is clearly unsuitable for transport studies. If (say) RNA appears in the medium in which such ruins are incubated in vitro, the process by which it has left a nucleus cannot be equated with transport; it is inherently more likely to be attributable to leakage through holes in the walls.

It is immediately obvious that this argument is founded on the presumption that nuclear RNA is soluble; that is, once again, on tacit acceptance of something like the 'glucose transport' perspective. Rejection of this perspective does not directly invalidate the criticism, but it removes its foundations and deprives it of much of its force. Interestingly, however, not only does the argument represent consensus opinion, it has also remained insensitive to substantive counters. For example, in vitro conditions have been devised in which mature messengers but not splicing intermediates are exported from nuclei,[36] which is difficult to reconcile with the idea that 'random leakage' occurs; and the egress of RNA is blocked by a known inhibitor of pore-complex-mediated transport, wheat germ agglutinin,[37] which is also difficult to reconcile with 'leakage' unless it is assumed that the lectin somehow repairs all the breaches in the nuclear envelopes. Results such as these suggest that nuclear RNA does not behave as a solute, and ipso facto lend support to something like a solid-state transport perspective.[33] It seems likely that their incompatibility with the accepted paradigm, the 'glucose transport' perspective and the tacit assumptions of in situ methodology, has some bearing on the fact that they are consistently ignored.

In vitro results consistent with random leakage can certainly be obtained, but only if at least one of the following conditions are met: (a) RNAase and proteinase activities are not inhibited or (b) the nuclei are allowed to swell.[24] The significance of condition (a) is immediately obvious (random fragments of nuclear macromolecules are solubilized), but (b) is more puzzling. Further tearing of the envelope when a nucleus swells could be a factor, but it is hard to see how this could make much difference if the envelope is torn already. An interesting possibility is that swelling perturbs intranuclear structures sufficiently to solubilize RNA; to peel the paper from the walls, so to speak. The fact that 'nuclear matrix' preparations can be obtained from unswollen[38] but not from swollen[39] nuclei may be relevant here, but the connection has not been demonstrated directly.

This is the paradox: the minimalist, reductionist perspective is firmly entwined in studies on intact cells, and the broad, holistic perspective in studies on isolated subcellular fractions. We do not wish to make exaggerated claims about the virtues of in vitro studies, many of which have been inadequately validated, and all of which must be artifactual at least in the sense that nuclear contents are being exported to an unstructured aqueous solution, not a structured cytoplasm. However, to reject them out of hand, sui generis, is to adopt the currently dominant paradigm uncritically, and we wish to avoid such prejudice. And it remains true that if any in vitro results are credible, then they are more likely to throw light on kinetic and energy-transduction considerations than any number of in situ data.

Another factor that may have contributed to the dismissal of the in vitro approach is its venerability. It is a product of the 1960s and 1970s and the newer, more technically sophisticated in situ approach has superseded it. The older work is not read and references to it are ignored. The cliché that scientists ignore papers more than 5 years old (except for one or two that are considered landmark publications in the field) is a cliché because it is largely true. If the practice is to be considered justifiable, then we must infer either: (a) that work more than 5 years old is irrelevant to current understanding, which is tantamount to saying that there is no such thing as a continuing tradition of scientific knowledge, and inevitably sets the 'landmark' publications firmly out of context or (b) that it is helpful and constructive for scientists to have limited historical perspectives and consequently restricted understanding of their fields. We are not comfortable with either inference, and therefore regard the cliché practice as unjustifiable. Again, this makes us willing to admit that work that is old is not necessarily ipso facto useless.

OVERVIEW AND PROSPECTS

Of course, we have not made our own perspective on nucleocytoplasmic transport clear. To have done so in our first chapter would have been to present a conclusion upon which our facts would subsequently be based. What we have done is to show from an analysis of our actual uses of the word 'transport', what perspectives we can *not* adopt. One of these, which we have dubbed analogically the 'glucose transport' perspective, has dominated our discussion because it represents the current viewpoint, the consensus paradigm. Our rebuttal of this position has been shown to entail a number of consequences. One is that the potential for significant linkages between nucleocytoplasmic transport research and other fields of cell biology is greater than might have been supposed. Another is that while some generally accepted methods need to be critically re-evaluated and their results reinterpreted, other less generally accepted methods might give valid data after all. A third, following the first, is that a range of aspects of cell biology (e.g., RNA metabolism, cytoskeletal and nucleoskeletal structure and function, dynamics of macromolecule movements within compartments) need to be considered along with studies on nuclear envelopes and the characteristics of transported molecules. And a fourth is that some other fundamental issues, such as the significance of 'ill (or well) characterized' and the validity of diffusion theory, need to be analyzed critically. This will be among the tasks of the next two chapters, and will involve us not only in further exercises in linguistic analysis, but also the challenging of some fundamental beliefs, which will probably cause dysphoria. It may stimulate some readers and arouse the wrath of others. So be it; what one perceives as the preparation of a feast, another might see as the slaughter of sacred cows, or as merely beefing.

Before we continue, let us repeat our earlier disclaimer. The rejection of one perspective does not automatically entail the adoption of another. Our skepticism has not been confined to the 'glucose transport' viewpoint, but has extended to the alternatives that have been proposed. Our wish, and our intention, is to proceed with the nearest humanly possible approximation to open minds.

REFERENCES

1. Alberts B, Bray D, Lewis J et al. Molecular Biology of the Cell. 2nd ed. New York: Garland, 1988.
2. Goodson HV, Kang SJ, Endow SA. Molecular phylogeny of the kinesin family of microtubule motor proteins. J Cell Sci 1994; 107:1875-1884.
3. Wieland FT, Gleason ML, Serafini TA et al. The rate of bulk flow from the endoplasmic-reticulum to the cell-surface. Cell 1987; 50:289-300.
4. Mizuno M, Singer SJ. A soluble secretory protein is first concentrated in the endoplasmic reticulum before transfer to the Golgi apparatus. Proc Natl Acad Sci USA 1993; 90:5732-5736.
5. Balch WE, McCaffery JM, Plutner H et al. Vesicular stomatitis-virus glycoprotein is sorted and concentrated during export from the endoplasmic-reticulum. Cell 1994; 125:239-252.
6. Rose JK, Bergman JE. Altered cytoplasmic domains affect intracellular transport of the vesicular stomatitis glycoprotein. Cell 1983; 34:513-524.
7. Franke WW. Structure, biochemistry and functions of the nuclear envelope. Int Rev Cytol Suppl 1974; 4:72-236.
8. Feldherr CM. The nuclear annuli as pathways for nucleocytoplasmic exchange. J Cell Biol 1962; 14:65-72.
9. Stevens BJ, Swift H. RNA transport from nucleus to cytoplasm in

Chironomus salivary glands. J Cell Biol 1966; 31:55-57.

10. Paine PL, Feldherr CM. Nucleocytoplasmic exchanges of macro-molecules. Exp Cell Res 1972; 74:81-98.

11. Paine PL, Moore LC, Horowitz SB. Nuclear envelope permeability. Nature 1975; 254:109-114.

12. Akey CW. Visualization of transport-related configurations of the nuclear pore transporter. Biophys J 1990; 58:341-355.

13. Agutter PS. Between Nucleus and Cytoplasm. London: Chapman and Hall, 1991.

14. Riedel N, Fasold H. Transport of ribosomal proteins and RNAs. In: Feldherr CM, ed. Nuclear Trafficking. San Diego: Academic Press, 1992:231-290.

15. Newmeyer DD, Forbes DJ. An N-Ethylmaleimide sensitive cytosolic factor necessary for nuclear protein import: requirement in signal-mediated binding to the nuclear pore. J Cell Biol 1990; 110:547-557.

16. Schröder HC, Bachmann M, Diehl-Seifert B et al. Transport of mRNA from nucleus to cytoplasm. Prog Nuc Acids Res Mol Biol 1987; 34:89-142.

17. Laurence JB, Singer RH, Marselle LM. Highly localized tracks of specific transcripts within interphase nuclei visualized by in situ hybridization. Cell 1989; 57:493-502.

18. Zachar A, Kramer J, Mims IP et al. Evidence for channeled diffusion of pre-mRNAs during nuclear RNA transport in metazoans. J Cell Bio 1993; 121:729-742.

19. Hesketh JE, Pryme IF. Interaction between mRNA, ribosomes and the cytoskeleton. Biochem J 1991; 277:1-10.

20. Singer RH. The cytoskeleton and mRNA localization. Curr Opin Cell Biol 1992; 4:15-19.

21. Feldherr CM. Ribosomal RNA synthesis and transport following disruption of the nuclear envelope. Cell Tissue Res 1980; 205:157-162.

22. Feldherr CM, Pomerantz J. Mechanism for the selection of nuclear polypeptides in *Xenopus* oocytes. J Cell Biol 1978; 78:168-175.

23. Paine PL. Diffusive and non-diffusive proteins in vivo. J Cell Biol 1984; 99:188s-195s.

24. Agutter PS. Nucleocytoplasmic transport of mRNA; its relationship to RNA metabolism, subcellular structures and other nucleocytoplasmic exchanges. Prog Molec Subcell Biol 1988; 10:15-96.

25. Zeitlin S, Parent A, Silverstein S et al. Pre-mRNA splicing and the nuclear matrix. Mol Cell Biol 1987; 7:111-120.

26. Bonner WM. Protein migration into nuclei I: frog oocyte nuclei in vivo accumulate microinjected histones, allow entry to small proteins and exclude large proteins. J Cell Biol 1975; 64:431-437.

27. Bonner WM. Protein migration into nuclei II: frog oocyte nuclei

accumulate a class of microinjected oocyte nuclear proteins and exclude a class of microinjected cytoplasmic proteins. J Cell Biol 1975; 64:421-430.

28. Agutter PS. Nucleocytoplasmic mRNA transport. Subcell Biochem 1984; 10:281-357.

29. Agutter PS. Models for solid-state transport: messenger RNA movement from nucleus to cytoplasm. Cell Biol Internat 1990; 18:849-858.

30. Kalderon D, Richardson WD, Markham AF et al. Sequence requirements for nuclear location of simian virus 40 large-T antigen. Nature 1984; 311:33-38.

31. Smith AE, Kalderon D, Roberts BL et al. The nuclear location signal. Proc Roy Soc B 1986; 226:43-58.

32. Dingwall C, Laskey RA. Protein input into the cell nucleus. Ann Rev Cell Biol 1986; 22:367-390.

33. Agutter PS. Nucleocytoplasmic mRNA transport; a plea for methodological dualism. Trends Cell Biol 1994b; 4:278-279.

34. Riedel N, Fasold H. Nuclear envelope vesicles as a model system to study nucleocytoplasmic transport. Biochem J 1987; 241:213-219.

35. Lang I, Scholz M, Peters R. Molecular mobility and nucleocytoplasmic flux in hepatoma cells. J Cell Biol 1986; 102:1183-1190.

36. Otegui C, Patterson RJ. RNA metabolism in isolated nuclei: processing and transport of immunoglobulin light chain sequences. Nucleic Acids Res 1981; 9:4676-4681.

37. Prochnow D, Thomson M, Schröder HC et al. Efflux of RNA from resealed nuclear envelope ghosts. Arch Biochem Biophys 1994; 312:579-587.

38. Berezney R, Coffey DS. Isolation of a nucler protein matrix. Biochem Biophys Res Commun 1974; 60:1410-1417.

39. Krachmarov CP, Iovcheva C, Dessev GN. A simple rapid method for isolation of nuclear lamina from Ehrlich ascites tumor-cells using DNAse-II. Dok Bolgar Akad Nauk 1984; 37:1423-1426.

CHAPTER 2

THE VAGARIES OF DIFFUSION

INTRODUCTION

In the first chapter we noted that molecule movements within a compartment are considered irrelevant to transport just as far as they can be attributed to diffusion; and conversely, so far as a molecule's movements are not due to diffusion, they must be considered part of transport. We offered prima facie reasons for doubting whether cytoplasmic or nucleoplasmic proteins and RNAs could be held to 'diffuse' over distances many times greater than molecular dimensions, but our contention rested on an intuitive understanding of that term. The issue is central to the question of how we should properly conceive of nucleocytoplasmic transport, and is highly relevant to other aspects of cell physiology. It bears directly on the interpretation of experimental work and on the direction of some theorizing. Therefore we need to address the notion of 'diffusion' in detail.

THE DESCRIPTIVE AND EXPLANATORY USES OF 'DIFFUSION' ARE NOT SHARPLY DISTINCT

'Diffusion' is a widely-used term, denoting the passive, spontaneous movement of something from regions where it is abundant to regions where it is sparse. It is often used qualitatively (descriptively), but it nevertheless conveys an inherently quantitative idea, that of a gradient between two spatial locations. Everyday language statements such as 'the scent of the flowers diffused slowly around the room' or 'the red wine diffused rapidly across the carpet' are unambiguous; in each case, 'diffused' describes a phenomenon but does not purport to quantify it precisely or to explain it. The physical chemist who measured the volume of wine spilt, determined by detailed measurements the rate of increase in wine concentration at several locations on the carpet, and then (after a calculation based on Fick's law or similar) declared that the word 'diffused' had been correctly or incorrectly applied, would properly be considered an intellectual sociopath and would have been better employed in cleaning up the mess. However, there is no doubt that the adverb 'slowly' in the first sentence implies that the air in the room was still and the olfactogen concentration was low even at source; perhaps also that the room was large. Likewise, 'rapidly' in the second sentence implies either that a large amount of wine was spilt at source (high concentration gradient), or that the carpet was particularly amenable to the spread of wine (high diffusivity), or both. The notions of concentration-time-distance relationships in an unstirred medium seem to be latent in our common usage of 'diffusion', and the two example sentences together suggest that we are willing to apply the term to both three- and two-dimensional systems.

The evident gray area between the descriptive (qualitative) and explanatory (quantitative or scientific) uses of the word entails a logical danger. It is difficult to be sure when one has crossed an unmarked or fuzzy border, and if it is crossed unawares, the fallacy known to medieval logicians as 'affirmation of the consequent' might occur. It is logically correct to deduce from: (1) 'In all systems obeying such-and-such a law of diffusion material moves from regions of high to regions of low concentration' and (2) 'S is a system obeying the aforesaid law of diffusion'; that (3) 'material in S moves from regions of high to regions of low concentration'. But it is fallacious to deduce from the same major premise and the experimentally-validated statement 'material in S moves from regions of high to regions of low concentration' that (3) 'S is a system obeying the aforesaid law of diffusion'. Put like this, the fallacy is obvious. It is akin to the McCarthyite argument 'All Commies hate my guts; Doe hates my guts: therefore Doe is a Commie'. But because our everyday use of 'diffusion' has such powerful quasi-quantitative connotations, the observed fact of migration 'down a concentration gradient', however unquantified, immediately evokes the concept and we intuitively align the system with the principles of diffusion theory. The fallacy can, and does, arise in practice. When intuitions are strong, fallacies are insidious.

Biologists who are asked informally whether the use of the key word in (say) 'respiratory gases enter and leave the early embryo by diffusion' is descriptive or explanatory almost all reply 'explanatory'. When challenged to provide evidence that the gas movements (approximately) obey Fick's law, they are unable to oblige. Thereupon some of them retract and declare the word to have been merely descriptive after all, while the rest inquire, often indignantly, what other explanation there can be since no specific biological process such as secretion is involved. The rightness or wrongness of attributing early-embryonic gas exchange to diffusion is not at issue here (though the question is not empty), but this highly reproducible scenario indicates that the fuzziness confounding description with explanation, and the danger of affirming the consequent, are far from being abstruse logical matters of no real relevance to science. On the contrary, these typical responses seem to corroborate the physicist's unkind judgment that biologists have concrete minds, in the sense that they are all mixed up and firmly set.

If (to pursue the same example) it is true that no biological process is involved in embryonic gas exchange and the process is therefore wholly explicable in physio-chemical terms, then it follows that the kinetics of exchange would be indistinguishable in an inanimate object identical in size, composition and organization to the embryo; for instance, a dead embryo that had undergone no cytolysis or other deterioration. Let us concede this for the sake of argument (it might even be true). Does it therefore *necessarily* follow that the 'physio-chemical terms' that are required come from diffusion theory? Obviously not. Diffusion theory might provide the appropriate explanation, but (logically, at least) it might not. However, the issue is not simply one of logic, but of reasonableness and need. 'What other explanation can there be?' is a fair question, or rather, two questions; to wit, 'Do we need a different explanation?' and 'If so, what is it?'. So far as macromolecule movements in cells are concerned, we need to address the first question, and perhaps, depending on the outcome of that investigation, the second. And to answer 'do we need a different explanation?' we have to know what the paradigm explanation, diffusion theory, has to offer, and whether it is reasonable to apply it to the conditions of the cell internum. This requires an analysis of the assumptions underlying diffusion theory itself.

CLASSICAL DIFFUSION THEORY INVOLVES AT LEAST TEN ASSUMPTIONS AND IS STRICTLY APPLICABLE ONLY TO IDEAL SOLUTIONS

Diffusion theory claims our credulity because it accounts for the macroscopic relationship between concentration gradient and rate of change of concentration through a given plane, first satisfactorily modeled by Fick,[1] in terms of the kinetic theory model of Brownian motion; or more precisely, in terms of the Maxwell-Boltzmann statistical dynamics of weakly-interacting particles showing unconstrained thermal motion.[2,3] It is widely believed that Fick's law was an empirical generalization from his laboratory data, but a study of his major paper[1] shows that his results had a high enough intrinsic error to be consistent with a wide range of mathematical patterns. In fact he proposed his law by analogy with Fourier's mathematical model of heat conduction through metal bars, which had been advanced three decades earlier and amply corroborated in the interim.[4] So much for positivism; but it can be taken as vindication of Fick's insight that both Fourier's law, and the Maxwell-Boltzmann statistics that were found to provide the explanation for his own law, are rooted in the principles of statistical mechanics. Both are contentious (see, for example, ref. 5). Some relevant aspects of this contention will become apparent in what follows. It is, in fact, possible to derive Fick's law from first principles[5] assuming random movements of particles in a liquid medium (following Einstein,[2] we use 'particles' to signify both suspended objects such as those in pollen grains and solute molecules in true solution). The complete derivation would be superfluous here (see Agutter et al[6] for a full account) but we need to provide sufficient detail to identify the five assumptions implicit in it. A further five assumptions (at least) underpin the derivation of the Einstein-Smoluchowski model of Brownian motion, to which we shall turn shortly.

The theoretical justification of Fick's law entails the demonstration that, if concentration change is measured along a particular coordinate, the rate of change across any plane perpendicular to that coordinate is proportional (as nearly as makes no difference) to the second partial derivative of the concentration gradient at the time of measurement. That is, if C = concentration, x = distance from the starting point in the specified direction and t = time:

$$\left(\frac{\delta C}{\delta t}\right)_x = D\left(\frac{\delta^2 C}{\delta x^2}\right)_t$$

where D, the diffusivity or diffusion constant, is equal to the square of the average distance migrated by the particles in time t divided by 2t:

$$D = \frac{\overline{x^2}}{2t}$$

The assumptions required are:

1. The time over which measurement is made is many orders of magnitude greater than the duration of the average Brownian motion. Fick implicitly assumed a continuous resistance to diffusion. This means ignoring van der Waal's and electrostatic interactions between particle and solvent[7] which is problematic if the distance (hence time) of migration is in the same order as the particle diameter. In this case, and indeed more generally if solvation shells behave chaotically, resistance and ipso facto the mean particle migration rate show discontinuities.

2. The solvent activity is constant at all planes through the direction of movement across which measurement takes place, i.e., the solution/suspension is homogeneous and infinitely dilute.

   If these two assumptions do not more or less coincide with the properties of the system, then the right hand side of the equation will not simplify to

$$\left(\frac{\delta^2 C}{\delta x^2}\right)_t$$

   but will contain numerically significant terms of the type

$$\left(\frac{\delta^{2n} C}{\delta x^{2n}}\right)_{t}, n > 1$$

3. Molecular bombardments of the particle by solvent are equally probable from all directions.

4. The system is unstirred (stationary). There is no bulk solvent flow; solvent molecules move only (i) by random thermal motion and (ii) in exchanging places with particles.

5. There are no significant particle-particle interactions, i.e., individual particles migrate independently.

Unless these assumptions approximate to reality, the requirement that each particle in the system is equally likely to move in any direction at any moment is not fulfilled, and the sum of the migration probability functions is accordingly not zero. This departure from ideality maintains the numerical significance of terms of the type

$$\left(\frac{\delta^{2n+1}C}{\delta x^{2n+1}}\right)_t$$

on the right-hand side of the equation. Therefore, if the system departs materially from any or all of these five assumptions, the direct proportionality between

$$\left(\frac{\delta C}{\delta t}\right)_x \text{ and } \left(\frac{\delta^2 C}{\delta x^2}\right)_t$$

is lost. The parameter D is then a function of concentration, location and time, and is certainly not a constant. In particular, significant particle-particle interactions lead to nonlinearities in the equation, precluding analytical solution, and sufficiently large bulk solvent flows swamp the effects of Brownian motion. For the present, we shall note that the validity of any application of Fick's law to the real world depends on how we interpret '*material*' departure from the assumptions, '*significant*' particle-particle interactions and '*sufficiently large*' solvent flows, and that we especially need to know how these interactions affect the application of diffusion theory to molecule movements in the cell internum.

Derivation of the Einstein-Smoluchowski model from the principles of kinetic theory requires the same five assumptions, some of which can be interpreted quantitatively in this new context. For instance, the 'high (infinite) dilution' of assumption 2 actually requires the particle concentration to be less than about

$$\frac{(3mkT)^{3/2}}{h^3}$$

where m = particle mass, T = absolute temperature, k = Boltzmann's constant and h = Planck's constant. The additional assumptions are:

6. All particles have the same mass, m.

7. The number of particles in any homogeneous part of the system is sufficiently large for statistical averaging to be valid. Since the particles must also be dilute (assumption 2), this requires large homogenous volumes (see also assumption 1).

8. Successive permitted translational energy levels are much less than kT. Generally, we would expect this assumption to hold unless there are strong particle-particle interactions (assumption 5).

9. When the function

$$\left(\frac{2RT}{mN}\right) e^{\omega t/m}$$

is integrated, the constant of integration A is negligible. (Here, R = the gas constant = Nk, where N = Avogadro's number, and ω is the damping coefficient on the particles due to viscosity). More precisely, we need to assume that $Ae^{-\omega t/m}$ is negligible, so if A is not close to zero then we require $\omega t >> m$; a high damping coefficient, long times of travel through kinetically homogeneous parts of the system (assumptions 1-2) and very low particle (molecule) mass. For the damping coefficient it is conventional, following Einstein again, to assume the Stokes relationship $\omega = 6\pi r \eta$, where r = particle radius and η = the coefficient of viscosity of the medium. Strictly speaking this requires that the particles be spherical, but in practice ω is not highly sensitive to particle shape.

10. The particles are rigid and particle-particle collisions are perfectly elastic (this is related to but not entirely equatable with assumption 5).

If these conditions are met then D is constant for constant thermodynamic temperature, viscosity and particle radius:

$$D = \frac{RT}{6\pi\eta N}$$

If they are not met then D is a complicated and probably analytically undecidable function of particle mass, homogeneous phase volume, location, time and concentration.

We need to remember that Einstein and Smoluchowski were concerned with the general validation of kinetic theory, and ipso facto atomic theory, and did not (need to) consider the applicability of their model to complicated or heterogeneous systems. Also, although Fick's interest was in physiology, he worked at a time when such attempts as had been made to provide mathematical accounts of the diffusion phenomenon were highly unsatisfactory. It should therefore not surprise us if the theory of diffusion proved inapplicable to real, highly complicated systems such as the cell internum. On the other hand, fault finding in diffusion theory and the quest for alternative models are inhibited not only by the linguistic-logical considerations we discussed earlier and by the sheer inherent difficulty of the task, but also by the conviction that Fick's law, like Fourier's, conforms to one of those 'general patterns of nature', like inverse-square relationships and first order kinetics, that gain credibility by their structural familiarity, and by the feeling that any challenge to diffusion theory seems (though in fact it does not) to question kinetic theory, or even atomic theory itself, to say nothing of the sanctity of Einstein's name.

Obviously the ten assumptions we have given break down in the context of the cell internum, but the question is, whether this really matters. It may be that diffusion theory applies to a sufficiently close approximation in the system we are addressing, or that it can be made to apply by dint of theory-based modification or the introduction of ad hoc adjustments. Therefore, an affirmative answer to the question 'do we need a different explanation?' is demonstrated only if we can show, as a matter of both practice and principle, that no amount of modification or ad hoc tinkering can make diffusion theory apply reasonably to macromolecules in cytoplasm or nucleoplasm. By analogy: frictionless surface-body interactions bear the same sort of relationship to real-world movements of objects across surfaces as ideal solutions bear to real ones, but the 'frictionless surface' assumption remains valuable for deducing mathematical accounts of observed mechanical behavior, at least sometimes. When Galileo rolled cannonballs down smooth inclined planes he had found a sufficiently close approximation to the ideal to permit experimental vindication of his theory of dynamics. (Galileo's capacity to deal in idealized abstractions suggests that his rejection of Aristotle made him in this respect, as in his commitment to a mathematically inscribed Book of Nature, a Platonist. In this way, as in so many others, his example seems to have shaped a good deal of later science). The ideal remains a reasonable approximation even when the plane and the cannonball surface are rough; and theory enables us to modify the equations to accommodate nonspherical cannonballs. Even when the topology of the moving object makes it depart from rectilinear motion, ad hoc adjustments can suffice to sustain the value of the model. However, if the surface is a succession of layers of sandpaper differing unpredictably in texture and gradient, and the moving object is an assembly of a dozen or so beans dipped in glue, then the initial momentum of the body ceases to be the primary determinant of motion, and other parameters, extrinsic to the theory, achieve major importance. In this case a fundamentally different model is needed. Likewise, if the moving body is a small lump of expanded polystyrene, its velocity at any instance is more likely to be determined by draughts than by initial momentum or surface characteristics. On the strength of this analogy, therefore, the nub of our problem seems to be: is the movement of a macromolecule over large distances within the cell dominated (usually) by Brownian motion? If 'yes', then modifications and adjustments of diffusion theory should suffice to explain it; if 'no', then in principle they cannot, and an alternative model involving other mechanisms is required. (The analogy we have used here functions unashamedly as justification not heuristic. We can therefore rely on it only to clarify, not to guide).

This general questioning of the use of diffusion theory in biology is not novel.[8-12] But the belief that diffusion theory is basically valid, that the cell internum belongs to its domain and all that is required to make the theory fit any possible experimental data is appropriate modification, is apparent in almost all publications about the movements of biomolecules (see, for example, refs. 13-16). Belief in diffusion theory is deeply entrenched in our thinking, which is why, in view of our overall purpose, we have devoted several pages to explaining the entrenchment. We are now in a position to tackle the central theme of this chapter: to consider whether 'diffusivity of an intracellular macromolecule' is a scientifically meaningful, or interpretable, or practically useful, expression when applied to movements over a substantial fraction of the cell's diameter.

LONG RANGE MACROMOLECULE MOVEMENTS IN CELLS ARE NOT DOMINATED BY BROWNIAN MOTION

A convenient starting-point is an assertion that will be accepted by practically all cell biologists: intracellular fluids certain high concentrations of macromolecules, which are sticky and are not rigid. This immediately divorces intracellular media from the constraints of assumptions 2, 5 and 10. The solution is not infinitely dilute, or anywhere close to that ideal; particle-particle interactions are frequent and often strong; and collisions between or amongst particles are seldom elastic, let alone perfectly so. It follows that solvent activity is not constant along concentration gradients, particles do not move independently, and the total kinetic energy of a given pair of particles is not conserved when they collide. The complexes formed when pairs or larger assemblies of particles interact might have life-expectancies many times greater than the duration of a single Brownian movement; indeed, shorter assembly lifetimes can hardly be expected in a system containing large numbers of sticky macromolecules per cubic micrometer.[17]

Several consequences follow. First, because there is a significant likelihood that metastable assemblies of macromolecules will be present in the cell at any instant, we must infer that there will be many such assemblies in (say) a given cubic micrometer of cytoplasm at any time. If the same volume is re-examined a second or two later most of these *particular* assemblies will no doubt have dissociated again, to be replaced by about the same number of others. Arguments about the 'sol' or 'gel' state of cytoplasm have an archaic ring, but it is clear that our agreed characterization of the medium (a highly concentrated solution of sticky particles) leads directly to the inference that it behaves as a dynamic gel. Electron microscope specimens of cells prepared under conditions likely to preserve the continuous solid phase of such a gel duly revealed a 'microtrabecular lattice',[18] an all-pervasive transcytoplasmic network with a mean lattice space of 10-20 nm. It is widely held that this result was misleading in that it does not signify the presence of a stable structure akin, say, to the cytoskeleton. No doubt this objection is valid; but the observation coincides with predictions from our inference that the medium is a *dynamic* gel. Unless the adjective is emphasized, a stable structure is connoted, and we agree that this is probably misleading. However, some assemblages, such as those representing functional sets of enzymes in metabolic pathways, may have longer lifespans than implied here.[19,20] If the 10-20 nm spacing is a valid measure, and it seems reproducible under preparation conditions that differ from one another in detail, then we must infer that the cytoplasm is kinetically homogeneous on a millisecond timescale only within volumes defined by the lattice, that is, over a few thousand cubic nanometers at most. Assumption 1, which implies large migration distances (and times) through a homogenous medium, is therefore inapplicable. Interactions between the particle and the ephemeral lattice itself might dominate movement; in any case, Fick's tacit assumption of continuous resistance is certainly invalid. The lattice must (however transiently) define its own internal gradients and is certainly solvated, so we must expect significant local fluctuations in solvent activity, further prejudicing the applicability of assumption 2.

Second, not only is the system heterogeneous in particle mass (since not all cellular macromolecules have the same M_r), but the formation of more-or-less random, more-or-less transient assemblies makes it more so. Assumption 6, that all particles are of equal mass, is very far from the reality of the cell. The consequence of this is that the impulse forces representing the total translational energy of each particle in the x direction

$$\left(m \frac{d2x}{dt2} + \omega \frac{dx}{dt} \right)$$

do not vanish when averaged over all particles in the system, as they do in the idealized model considered by Einstein[2] and Smoluchowski.[3] Since the derivation of the statement

$$\frac{d(\overline{x^2})}{dt} = \int \frac{2RT}{mN} e^{\omega t/m} dt ,$$

on which the inference

$$D = \frac{RT}{6\pi\eta N}$$

is based, depends (a) on this mutual canceling of these impulse forces and (b) on the particle concentration being low enough

(less than $\quad\dfrac{(3mkT)^{3/2}}{h^3}\quad$)

for the Equipartition of Energy Theorem to be applied, it follows that an intracellular D must be an unguessably complicated function of time, distance, concentration and no doubt other parameters, and cannot be constant. The canceling of the impulse forces also depends on the presence of large numbers of similar-mass particles in the homogeneous subcompartment (assumption 7), and it therefore cannot be expected, or assumed, that bombardments of the particle with solvent molecules are equally probable from every direction. The variability of solvent activity argues against this assumption (assumption 3). The principle by which Brownian motion is linked to macroscopic diffusion therefore has no place in an account of processes in the cell internum.

These arguments are enough to suggest that macromolecule movements inside a cell are no more amenable to description by classical diffusion theory than the rolling of glued beans over sandpaper is to a discourse in terms of frictionless body-surface interactions. However, more difficulties should be added. It is hard to know what sense to make of bulk parameters such as the coefficient of viscosity (η) and thermodynamic temperature (T) in the context of cytoplasmic subcompartments in the order of only 10^3-10^4 nm^3.[8,12,21] Accordingly, it is difficult to interpret ω, except to say that it is locally extremely variable; and assumption 9 ($\omega t \gg m$) cannot be justified when ω is uninterpretable, t very small and m variable but often quite large. 'Intracellular viscosities' are normally computed from experimentally measured 'diffusivities', using

$$D = \frac{RT}{6\pi\eta N}$$

as if its validity were assured. Assumption 8 (intervals between translational energy levels <<KT) may be valid in principle, notwithstanding the difficulty concerning T, but the translation events

envisaged in Maxwell-Boltzmann statistics are obviously swamped by the tendencies of particles to interact and form assemblages. And as for the one surviving assumption, number 4, bulk solvent flow definitely does occur inside cells. Apart from osmotic flux across the cell surface, where the permeability coefficient of water is reckoned as high as 10^{-4} m s^{-1}, there is: (a) cytoplasmic streaming, which may now be quantifiable in some systems;[22,23] (b) metabolic production and utilization of water leading to local fluxes around enzyme active sites; (c) the effects of particle assembly-disassembly process on the local chemical potential of water and (d) the effects of asymmetric exchanges (e.g., when a large or highly-hydrated particle changes place with a small or poorly-hydrated one, there will be a net movement of water molecules to fill the space). Water inside organisms is stationary, if ever, only when the organism is dead.[10,24]

Enormous theoretical difficulties therefore confront anyone who wishes to assume that intracellular macromolecules are 'diffusive' in the sense entailed in diffusion theory. (Whether this sense itself bears detailed scrutiny is open to doubt; for instance, Tyrrell[25] has raised the question of whether diffusivity, D, measures an intrinsic property of the particle in the system, or the exchange rates between particles and solvent molecules. Either interpretation seems consistent with classical diffusion theory). Can the theory be adapted or modified to suit the case? Certainly diffusion theory has advanced well beyond its classical beginnings and has variants that fit it to heterogeneous and mixed-phase systems, but not of the complex and dynamic type encountered in cells, where the half-lives of many relevant molecular assemblies are in the same order as the local half-times of particle migration (see, for example, Crank[26]). The various ad hoc modifications introduced in particular situations are not generalizable.[27-29] Attempts to replace the kinetic-theory formulation with a statistical mechanical treatment[30] are vitiated by the mathematical complications involved, which are fairly severe even for 'ideal' solutions; there seems no prospect of a credible or useful statistical mechanical model of macromolecule movements in cytoplasm.[31] It therefore seems that the theory cannot be suited to the case, and the difficulties are not just enormous, but insuperable.

These comments may seem bleak, but the inescapable conclusion that macromolecules in cells cannot be considered diffusive is on the one hand stimulating (a wholly new model is needed and all reasonable offers will be considered) and on the other unsurprising. It has long been obvious that diffusion cannot be the only kinetic mechanism for intracellular transport, otherwise molecule distributions inside cells would tend towards thermodynamic equilibrium rather than steady state;[32] so there has always been a need to find fundamentally different principles. The only difference that our conclusion makes is that these alternative principles need to *replace*, not to supplement, diffusion theory. And that in turn should not surprise us, because another belief in which most biologists would concur is that the insides of cells are highly organized down to the level of the molecule, and it would be paradoxical if a fundamentally random process such as diffusion played a large part in maintaining high order amongst the very entities whose distribution is randomized.

Intracellular macromolecule movements are governed by their interactions with each other and with organized structures, both stable and transient, within the system. In general, the only likely effects of Brownian motion are to introduce some uncertainty into the precise location of a given macromolecule at any moment, perhaps within a few tens of nanometers. What Brownian motion does not do in the cell, so our assessment of the theory indicates, is to afford the principal means whereby molecules move in bulk or over large distances, either down concentration gradients or (perhaps on occasion) otherwise. In general, cellular macromolecules do not migrate because of Brownian motion; they are not diffusive.

MEASUREMENTS OF INTRACELLULAR 'DIFFUSIVITIES' REQUIRE RADICALLY NEW INTERPRETATIONS

A child repeatedly warned about the dangers of hot stoves invariably burns its fingers, and biophysicists publish values for intracellular 'diffusivities'. That is to say, they microinject known volumes of known concentrations of the test substance into a cell, and determine its subsequent distribution by autoradiography (usually ultralow temperature), fluorescence microscopy, microphotolysis, or some other technique.[33-35] A graph of the rate of change of concentration at each location against the second derivative of the concentration gradient can usually be fitted to a straight line (not least because of data variance) and the gradient of the line is called 'diffusivity'. In view of our discussion about the concept of diffusivity as applied to intracellular media, it would be simpler and less misleading to present the concentration-time dependences directly. The data are valuable, because whatever model we finally accept for macromolecule movements in cells, it will need to be assessed against just such experimental observations. But the usual interpretations, including 'diffusivity' values, are quite another matter.

A convenient and critical overview of these data was provided about a decade ago by Peters and his colleagues.[36,37] Artificial molecules such as dextrans are useful probes for intracellular migration processes because they are unlikely to form complexes with or adsorb to endogenous molecules and structures. In general, their 'diffusivities' are around 6-8 times lower than their diffusivities in ordinary aqueous solution, and because a protein of M_r 25,000 shows a 50% mobility restriction in Sepharose beads with a 50 nm pore diameter,[34] this is broadly consistent with a 10-20 nm lattice spacing in the cytoplasm (see earlier discussion). Although the consistency with electron micrographs of the cytoplasmic gel[18] is striking, it might be no more than coincidental. The conditions of measurement need to be fully specified, because we expect 'diffusivity' in the cell to depend inter alia on concentration and location. At least, under constant conditions of measurement, dextran 'diffusivities' are directly proportional to temperature, as we would expect from the 'Stokes-Einstein' equation, so interpretation of the data in terms of obstruction to movement is reasonable. Unfortunately, the available techniques do not allow fine discrimination; even fluorescence microphotolysis, which gives the best resolution of all current methods, has limits of about 1 μm in distance and 1 ms in time.[36] This does not allow small-scale asymmetries to be observed, or even larger-scale anisotropies, because the results are averaged over different locations a given distance from the starting point. The averaging of no doubt distinct local behaviors over large regions of the cell probably accounts for the typically high variance in the data and the illusion of a constant 'diffusivity' value.

In contrast, the 'diffusivities' of proteins and RNAs measured by these methods are much lower than those calculated in free solution, often by more than two orders of magnitude. The contrast with dextrans is presumably attributable to binding, adsorption and complex formation, and (whatever objections there are in principle to averaging over a heterogeneous population of migration events) this is a striking corroboration of our conclusion from theory that Brownian motion makes only a trivial contribution to intracellular macromolecule movements. Moreover, the measured 'diffusivities' vary neither with temperature (except perhaps negatively) nor with M_r. The Einstein-Smoluchowski model predicts that they should vary directly with the former parameter and inversely with the cube root of the latter (since particle radius is more or less proportional to the cube root of M_r). The inadequacy of diffusion theory to account for intracellular macromolecule movements is therefore not only theoretically established, but experimentally corroborated.

Horowitz and his colleagues, who used ultra low temperature autoradiography to measure the intracellular diffusivities of low M_r solutes, have reported D values lower than those in 'free' water for uncharged substances such as sucrose, and have presented evidence that charged substances such as ATP bind extensively to intracellular structures.[38-40] These findings are generally accepted, and in view of them, the results (above) concerning intracellular dextran, protein and RNA mobilities are hardly surprising. This group has also investigated intracellular distributions by cryomicrodissection (separation of the nucleus from the cytoplasm of cells that have been flash-frozen in liquid nitrogen), and has extended this technique by use of an internal reference phase. The internal reference phase is a gelatin solution, of which a volume roughly equal to that of the nucleus is injected into an amphibian oocyte. After 20-30 hr, during which any diffusive cytoplasmic protein would be expected to equilibrate with the reference phase, the three subcompartments (gelatin, cytoplasm and nucleus) are separated by cryomicrodissection and their protein contents compared by two-dimensional gel electrophoresis.[41] The protein composition of the nucleus is markedly different not only from that of the cytoplasm, but from that of bulk-isolated nuclei. Significantly, only about 12% of the separately identifiable cytoplasmic proteins are present in the reference phase, supporting earlier observations that most cytoplasmic proteins are not diffusive.[42] Again, experimental findings coincide with the theoretical argument that intracellular proteins are not, for the most part, subject to diffusion.

If the nucleus loses most of its proteins during bulk isolation

that nuclei do not behave as osmometers), then the simplest inference is that nucleoplasmic proteins (in contrast to cytoplasmic ones)
are predominantly soluble. However, when the nuclear envelope is
torn open with a microneedle in situ, there is no measurable redistribution of nuclear proteins over a period of hours.[43] A more
drastic alteration of the system yields a different result. Dreyer et
al[44] removed the entire envelope from an oocyte nucleus in situ,
and found that nuclear proteins became evenly distributed over
the cell in a period of 10-30 min. Although this seems incompatible with the hypothesis that nuclear proteins are soluble in situ (a
soluble protein should spread much faster), we are led by these
various findings to the conclusion that a nucleus loses most of its
proteins when it is either torn from its biological habitat or, alternatively, flayed alive, but that it retains its proteinaceous entrails
when it is merely wounded by a massive incision. How this heterogenous set of results should be interpreted is a problem to which
we shall return at the end of this chapter.

If the measured 'diffusivities' of intracellular macromolecules
do not relate to a diffusion process, as they evidently do not, how
should they be interpreted? The observations we have summarized
here suggest that adsorption to stable intracellular structures, as
well as complex formation of the type suggested by the 'microtrabecular lattice', could play a part in directing and controlling
migration. As it stands, this is merely a sketch of a model; but
even in this embryonic form, it seems closer to reality than any
explanation based on diffusion theory. An adequate model should
be able to account for the temperature and mass independence of
protein 'diffusivity' reported by Lang et al[37] and for the reluctance
of most proteins to enter a reference phase.[41] If the model also
provided a credible explanation for the apparently conditional
mobility of nuclear proteins (see above), then it would have a reasonable claim on our attention. The challenge now is to establish
an alternative to a diffusion based model that meets these empirical constraints.

A MODEL FOR MACROMOLECULE MOVEMENT WITHIN COMPARTMENTS MIGHT ALSO BE APPLICABLE TO NUCLEOCYTOPLASMIC TRANSPORT

The behavior of nuclear proteins described above is reminis-
cent of the well-known fact that however immobile a cell's proteins

are in situ, a fair percentage of them become soluble when a cell is lysed or homogenized. This resemblance should not mislead us into supposing that there is only one type of solubilization mechanism for all those cellular proteins that *can* become soluble if the cell is sufficiently abused; but neither should it be overlooked. Various hypotheses can be suggested. For example, if a protein binds to sites in the cell with low affinity but the number of binding sites is high, then solubilization is easily explained.

We stress that this is not the only possible explanation (obviously the hypothesis does not account for the solubilization of tubulin or actin in tissue homogenates), but we have a particular motive for focusing on it here. The idea is elementary. Assuming the simplest model, $P + R \Leftrightarrow C$ (where P = free protein concentration, R = free binding site concentration and C = concentration of complex), so that $K_d = PR/C$, the ratio of unbound to bound protein = $P/C = K_d/R = K_d/(R_t-C)$, where R_t = the total concentration of binding sites. When the system is diluted, as when the cell is homogenized, (R_t-C) decreases while K_d presumably remains constant, so P/C rises. Also, the greater the value of R_t, the smaller the ratio P/C. Even if K_d is rather large, a high R_t value will ensure that most of the protein is bound. If we express the values of all the parameters in units of molecules per cell (for a cell with liquid contents of 10 pl, 10^7 molecules per cell is in the order of 1 μM), percentages of free protein for a range of R_t values are shown in Table 2.1, where a K_d of 10^6 molecules per cell (around 0.1 μM) is assumed. This example of minimalist kinetics is simplistic in several ways. We have no good reason to suppose that the binding sites form a kinetically homogenous population, or that the size of that population is constant, or that a simple hyperbolic saturation curve would be observed. Our use of the word concentration is glib (what does the term signify when the binding sites are ex hypotheso insoluble?), though we have given values in molecules per cell in Table 2.1 as a partial answer to this objection. (Parenthetically, it is striking that biochemists can happily use the word 'kinetic' to describe a model that implies no net movement of anything). But Table 2.1 serves to make the point that even with a rather low affinity, all but the highest abundance proteins are very largely immobilized in situ if the number of binding sites is sufficiently high. A more sophisticated model would make the same point, but perhaps less obviously.

Table 2.1. Predicted percentages of an intracellular protein that are free as opposed to bound, assuming a homogeneous population of binding sites with $K_d = 10^6$ molecules per cell (approx $10^{-7}M$) and simple hyperbolic binding

R_t		8	7	6	5	4
	8	7	90	95	100	100
	7	1	37	91	100	100
P	6	1	9	62	99	100
	5	0	0	1	95	100
	4	0	0	0	10	100

Key: R_t = log number of binding sites in cell, P = protein abundance (log number of molecules per cell).

How this model can be related to macromolecule transport within the cell is the next problem, but before we address it directly, we should explain our motive. Proteins scheduled for nuclear import bind to their cytoplasmic and nuclear envelope receptors with affinity constants in the order of $10^7 M$,[45,46] and similar nuclear envelope binding affinities have been reported for tRNAs and mRNAs.[47-49] This value coincides with the K_d assumed in Table 2.1. If (contrary to fact) we suppose all binding sites for a nucleocytoplasmic transport substrate to be confined to the pore-complexes, then presumably the receptor number is in the order of 10^4; there are usually a few thousand pore-complexes per nucleus[50] and it is difficult to imagine more than a very few simultaneously functioning docking sites for a given transportant per pore-complex. If proteins (or the complexes containing nuclear proteins; see chapter 4) were as mobile in the cytoplasm as glucose is in unstructured aqueous solution, then this would be unproblematic. A moderately abundant protein or complex of, say 10^5 copies per cell would give an initial 95% receptor occupancy, and if the protein crossed the nuclear envelope at (say) 1 molecule per pore complex per second when a receptor was occupied, then the initial entry rate would be 9.5×10^3 molecules per second and the time for half-complete loading of the 10^5 molecules into the nucleus would be around 13-14 s. However, the actual limit on cytoplasmic protein 'diffusiveness' suggests that it would take a typical protein 20-30 min to cross an average cell diameter.[51-53] If we suppose the distance between the translation site and the nearest point on the nuclear envelope to be a fifth of the cell diameter, the transit time to the nearest pore complex should then be in the order of 4-6 min. The time to reach any given pore-complex is presumably $\mu(\sqrt{d^2 + 2dr} + \phi r)$, where μ = the mean migration rate of the protein, d is the shortest distance between the translation site and the nuclear surface, r is the nuclear radius and ϕ is the radian angle subtended at the center of the nucleus between a tangent to the nuclear surface from the translation site and the target pore-complex (Fig. 2.1). If d = 6 μm, μ = 1 μm min^{-1} (above), and r = 5 μm, then the maximum time for arrival at a target pore-complex (at the opposite pole of the nucleus from the translation site) is 9.8 + 5ϕ min (the distance is 9.8 + 5ϕ μm), where ϕ = π - $\cos^{-1}$ (r/(d + r)) $\approx$ 2, giving a value of 19.8 min. If we further assume a 'normal' distribution of arrival frequencies at pore-complexes, of the type

$$F = \frac{1}{\sqrt{2\pi}} \int_{l_{max}}^{l_{min}} e^{-\frac{1}{2}l} \, dl,$$

where l is the distance traveled and l_{min} = 6 μm, l_{max} = 9.8 + 5 (π - $\cos^{-1}$ (r/(d + r))) μm, then the mean frequency occurs at around 10 μm, i.e., in 10 min, suggesting a mean arrival rate of proteins at the pore complex of 0.10 min^{-1} $\approx$ 0.002 s^{-1}. Because this is so much slower than translocation into the nucleus, notwithstanding the low receptor affinity, the period of nuclear loading will be more or less linear and therefore the process will be half complete in around (0.5 x 10^5)/(0.002 x 10^4) $\approx$ 2500 s, or about 40-45 min. This is not a viable rate. It follows that proteins or their transport complexes are presented to the pore-complex in a more efficient manner. It also follows as a matter of logic that transportant receptors are not confined to the pore-complex; this inference coincides with experimental evidence (see chapter 4).

Doubtless several hypothetical solutions to this problem can be advanced, but before we consider any of them it is worth calling

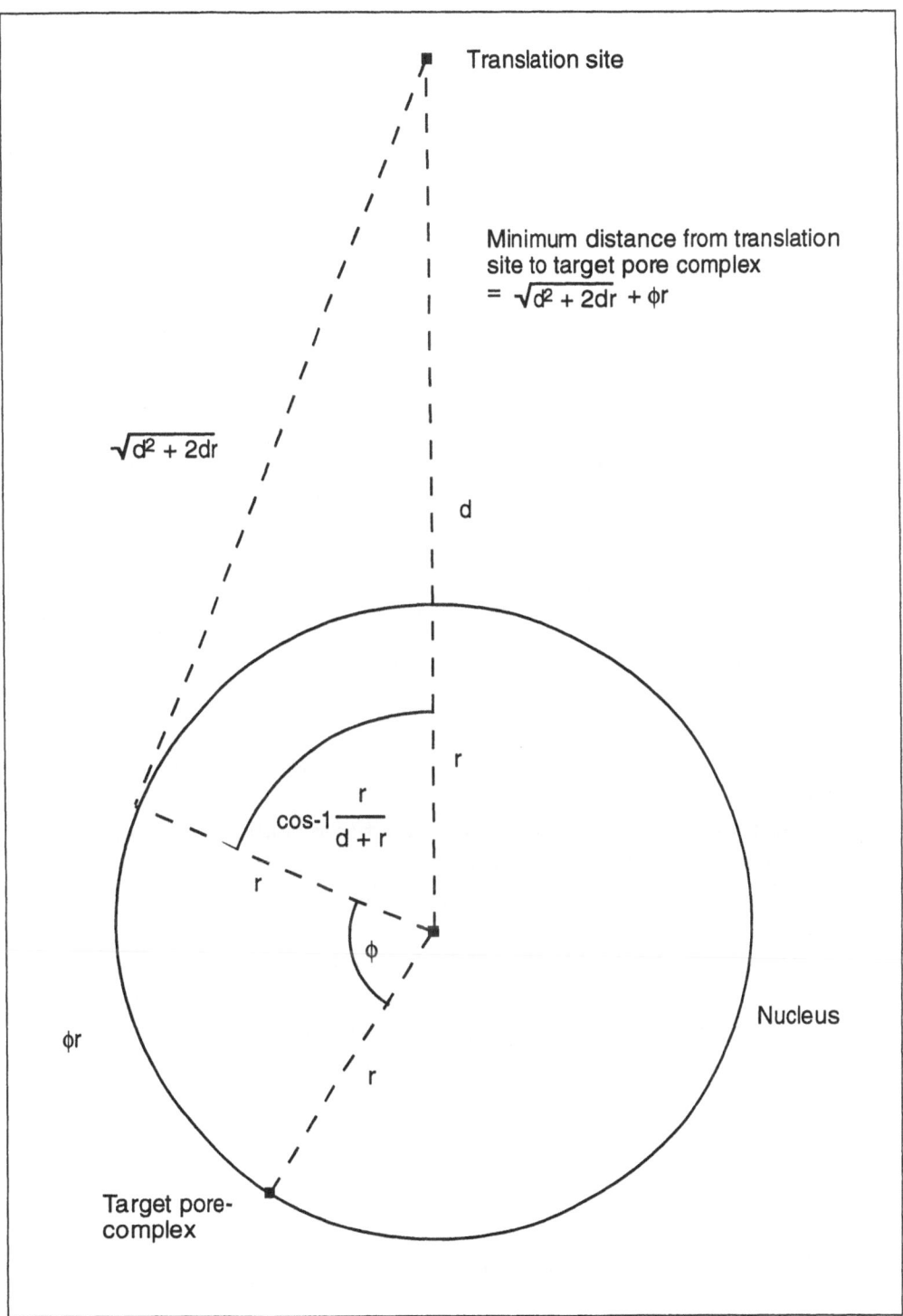

Fig. 2.1.

diffusivities' are so low, the focus of interest in nucleocytoplasmic transport must shift from the pore-complex to transport events in the cytoplasm itself, as a result of this simple mathematical reasoning from established data. This is not to say that events at the pore-complex are not worth exploring, of course; but if we want to understand the complete mechanism of nuclear protein uptake, we need to focus on events in the cytoplasm not at the pore-complex; events of the same general kind, we would presume, as any other kind of macromolecule migration in the cytoplasm. By analogy, this reasoning makes us more willing to believe (though obviously it does not demonstrate) that the kinetically interesting events in nuclear RNA export might take place within the nucleus, not at its periphery. This is why we suggested earlier in this book that the study of nucleocytoplasmic transport overlaps with the more general study of macromolecule movements inside cells, and that exclusive attention to the nuclear envelope is an error, because it implies the validity of a theory (diffusion) which we have shown is fundamentally inapplicable to the cellular milieu.

VIABLE MODELS FOR MACROMOLECULE TRANSPORT MUST INVOLVE INTRACELLULAR STRUCTURES OR FLUID MOVEMENT

Low protein 'diffusivities' in cytoplasm have previously been modeled in terms of 'adsorption chromatography' superimposed on diffusion. The cell contains R_t low-affinity binding sites of dissociation constant K_d, which have the effect of decreasing the 'diffusivity' of the protein in the fluid phase by a factor of $(1 + (R_t/K_d))^{-1}$ (see, for example, refs. 38, 51, 54). Apart from the incoherence of 'diffusivity' as applied to cytoplasm, this model has the defect that R_t is generally in the same order as the number of protein molecules, so given that the cytoplasmic 'diffusivity' is generally $\leq 0.01 D$ as measured in aqueous solution, it follows that $R_t \geq 100\, K_d$ and hence $R_t \geq 10^8$ given a K_d of 10^6 molecules per cell; and as Table 2.1 shows, 10^8 binding sites with $K_d = 10^6$ effectively immobilizes proteins. Of course, when a protein is microinjected into a cell, its concentration greatly exceeds R_t, so the chromatography analog becomes credible; but few proteins are present endogenously in more than 10^8 copies per cell (this is about the number of core histone molecules in a diploid human cell), and when protein content = $R_t = 10^8$ and $K_d = 10^6$, only about 7%

of the protein is unbound (Table 2.1). This is one reason why microinjection experiments can mislead our thinking.

Another way of putting this is to say that an individual endogenous protein molecule spends 93% of its migration time bound to (at least temporarily) immobile structures during transit. If the overall average transit speed is to exceed 1 μm min^{-1}, the value that led to the delayed nuclear uptake inference in our previous scenario, then the protein's average speed between binding sites must accordingly exceed 14 μm min^{-1}, and for a protein of only moderate abundance, say 10^5 copies per cell, it must exceed about 50 mm min^{-1}. This is credible only if the protein is transported between binding sites by rapid fluid movement with a high net velocity in the appropriate direction, which means cytoplasmic streaming directed to meet the needs of cellular transport. Mechanisms of cytoplasmic streaming have been investigated,[55] but so far as we are aware no one has indicated how it might be operated and directed to meet the cell's moment-by-moment protein transport needs. Very rapid streaming must be envisaged, because our constraint (net migration rate to exceed 1 μm min^{-1}) is a minimal one, and in reality proteins cross cells much more rapidly than this; [36] but it is far from certain that the cytoplasm is a landscape of rushing torrents (though see refs. 9, 56, 57 for a discussion of models of this kind, and Coulson[57] for an argument that metabolic processes at least in multicellular organisms are governed by fluid flow rates). Of course, binding sites for some proteins may be fewer and of lower affinity than we have supposed, but to judge from results such as those of Paine[41] such proteins are in a fairly small minority. Finally, although independence of M_r might be expected in this model, temperature independence is difficult to argue. The mechanisms of flow generation are insufficiently understood. Even the independence of M_r is dubious, because a liquid current might carry large and small molecules at different rates. If proteins are not transported by shooting intracellular rapids, that is if the rapids are not rapid enough, then the only logical alternative is that they move between binding sites without (significantly) entering the solution phase at all. Suppose, arbitrarily and for simplicity, there is a rectilinear array of identical binding sites (K_d = 10^6 molecules per cell) each 5 nm from the next. To exceed an overall transit rate of 1 μm min^{-1}, the time of protein transfer from one site to the next must be less than 0.3 s, which is well within the bounds of possibility. After all, the rates of metabolite transfer between contiguous enzymes in a 'metabolon' are much faster than this (see papers in ref. 58 for discussions of this). On this assumption, 10^8 binding sites should occupy a total length of 0.5 m, enough to cross the cell several thousand times, which is also credible because transcellular protein migration cannot generally be supposed to involve just one or two discrete and isolated trade-routes. The model might also explain the temperature-independence of protein migration rates[37] (see chapter 7). Independence of M_r is easily understood, because there is no reason in this model to suppose that M_r should be relevant to the transport rate. The slow (often infinitely slow) entry of an internal reference phase is also explicable by this model.

Its drawback lies in its arbitrary and simplisitic presuppositions. The arbitrariness of our proposed 5 nm spacing is a trivial matter, but the number of *identical* binding sites entails the involvement of a major cellular protein ($R_t = 10^8$ copies per cell), and if the arrays (which we have envisaged as rectilinear but need not be) are to function efficiently in transport they must be fairly stable. On the face of it, therefore, this apparently attractive model points to the involvement of the cytoskeleton in transcytoplasmic protein transport, and a fortiori in protein import by the nucleus. This inference cannot be rejected on logical grounds, but surely it is not a necessary consequence of the model, or we should be forced to conclude that the cytoskeleton has a direct though perhaps passive role in the transport of all 'soluble' proteins in the cell. This is feasible, but nonetheless difficult to believe. An escape route is to drop the assumption that the binding sites are all identical. This makes the kinetics so complex that experimental elucidation seems impossible. Generally, a transfer step from the ith protein-site complex (C_i) to the jth site (R_j) would be of the form

$$C_i + R_j \underset{K_{-j}}{\overset{K_{+j}}{\rightleftharpoons}} C_j + R_i, \quad K_{ij} = \frac{K_{-j}}{K_{+j}}$$

and the rate of occupation of the nth (final) site, under steady-state assumptions, is then

$$\frac{dC_n}{dt} = \frac{1}{R_1} \prod_{(j=i+1)}^{n} K_{ij} \quad \frac{dP}{dt}$$

where dP/dt represents the rate of loading of protein on to the first receptor, R, which in one interpretation of 'steady state' might be equal to the translation rate. If the array is not stable or fixed, the processes represented here by single rate-constants $K_{\pm j}$ are potentially more complicated, and will generally include not just the transfer event but the formation of a $C_i R_j$ pair, say with a quasi-Michaelis constant M_{ij}. The system is then described by the recursive equation

$$\frac{dC_j}{dt} = \frac{(K_{+j} - K_{-j})\, C_i\, R_j}{M_{ij} + C_i}$$

and a high M_{ij} could account for slow reference-phase penetration by some proteins.

Despite the difficulties of practical interpretation, this nonfixed system of heterogeneous binding sites may be a more realistic scenario than its simple predecessor. But in biological terms, it is not a generalized version of the simplistic model; it is a different model. The simple idea directly involved the cytoskeleton, but the more complicated one does not, except that the transient protein assemblies might be 'seeded' and briefly stabilized by the cytoskeleton.[18] What appears to be a single model for protein transport is, in fact, two models, each of which might be applicable to some but not all macromolecules in the cytoplasm. The stable homogeneous array invites comparison with other possible cytoskeleton-dependent mechanisms, such as motor-driven movement of occupied receptors analogous to axonal transport, or transfer of receptors resulting from dynamic processes such as treadmilling within the fibril itself.[59,60] Cytoplasmic streaming is unlikely to be relevant to any of these. In contrast, the unstable heterogeneous array may well be affected by cytoplasmic streaming. Fluid movements could accelerate or retard the formation of a $C_i R_j$ pair, especially if one member of the pair were more mobile than the other in the liquid phase. This could, in turn, entail indirect dependence on the cytoskeleton, which is implicated in at least some cytoplasmic streaming processes.[55] We might add that, in general, we would expect cytoskeletal dynamics and ipso facto any transport process directly dependent thereon to be temperature-sensitive.

Any of these models (transport by cytoplasmic streaming, by direct transfer between identical or heterogenous binding sites, by a mechanism akin to axonal transport, or as a consequence of processes such as treadmilling) can in principle account for rapid protein movements in the cell. Directly or indirectly, all of them involve the cytoskeleton, or intracellular fluid movements, or both. No doubt many alternative models could be proposed, depending on the ingenuity of the proponent. One further alternative, however, merits particular attention. Notwithstanding our demonstration that intracellular macromolecules cannot be said to 'diffuse', is it possible that in some cases migration can be attributed to a process kinetically akin to diffusion within channels defined by structural elements in the cell such as the cytoskeleton? Old intuitions die hard, and if the measured transport rate of any macromolecule over some region of the cell is sufficiently high, then the transport kinetics could resemble those of diffusion sufficiently to inspire disinterment of the theory.

The idea of 'diffusion channels' was proposed as a general metazoan phenomenon in a careful and detailed study of Zachar et al,[61] not in relation to protein movements in the cytoplasm, but to mRNA movements in the nucleus. We shall return to this paper in later chapters, but their account of mRNA migration needs to be considered here. First, it should be noted that the experimental model chosen by these authors gave unusually clear micrographs. They examined the large, uncluttered polytene nuclei of *Drosophila* salivary glands, and studied the movement of transcripts of a chimeric gene by fluorescence microscopy. One intron of this transcript is spliced cotranscriptionally, and another very slowly, so that the nuclear retention time is long. Second, they observed that the transcripts formed a web-like transnuclear array, and their evidence suggested that this array was defined not by nucleoskeletal fibrils, but by zones of exclusion from the chromatin. Third, their findings were in general accord with others (e.g., Jackson et al[62]) showing that hundreds of copies of particular transcription/processing sites are often distributed apparently at random across the nucleus. The inference drawn by Zachar et al[61] was that in many, perhaps most, cases, transnuclear migration of mRNAs is not specifically or anisotropically directed by the nucleoskeleton, but is a matter of isotropic, fluid phase movement through the interstices of the 'extrachromosomal network' (as they dubbed their system of channels). Implicitly, immature messengers are confined to the nucleus by the permeability limitations of the pore-complex. The attribution of intranuclear migration to 'diffusion' seemed difficult to resist.

On the one hand, it is not difficult to refute their hypothesis as it stands. Their conclusion that migration accorded kinetically with diffusion was based on the movement rates of microinjected dextrans and other inert particles,[36,63] which as we have already seen are inadequate models for this extrapolation. The total volume of the extrachromosomal network is about one tenth of the nuclear volume, implying a tenfold concentration of 'soluble' nuclear components, which must at least result in extensive adsorption of splicing factors and of the HnRNP itself to insoluble elements. The mean channel width of the extrachromosomal network is scarcely greater than the radius of gyration of HnRNP, which in itself rules out diffusion as a mechanism.[53] Also, the authors excluded the alternative solid-state transport hypothesis on the grounds that it necessarily implies the mechanism, and the migration rate, of fast axonal transport, which is an unnecessary and inappropriate limitation.[60] 'Channeled diffusion' is simply not a credible mechanism, and Zachar et al[61] set up only one rather improbable version of the alternative perspective as a straw man to attack. On the other hand their data were of exceptional quality, and there can be little doubt that they showed that nuclear RNA migration is not necessarily directly linked to a stable, rigid nucleoskeleton. It is possible that the results can be explained by net intranuclear fluid movements (which in this case might not need to be very rapid, given the slow overall migration rate), or by site-to-site transfer of bound material with some degree of desorption, presumably with the low-affinity binding sites on the nucleoskeleton; in other words, by one or other of the models we have already discussed. But these interpretations are conjectural, and the data serve to remind us that the meaning of 'nucleocytoplasmic transport' is far from clear as yet.

There is another reason for our detailed discussion of this particular paper in the present chapter: it suggests an explanation for the problematic behavior of 'soluble' nuclear proteins mentioned earlier. Suppose some nuclear proteins are, within limits, mobile in the fluid phase in situ, and the fluid phase is confined to a narrow extrachromosomal network. An incision in the nuclear envelope will connect only a very few channels of this network with the cytoplasm, so protein egress will be extremely and perhaps immeasurably slow. Complete removal of the envelope, however, will open a considerable number of channels, allowing the proteins to escape at a measurable rate. When nuclei are isolated in bulk using aqueous media, the long operation times, the extensive envelope tearing and the exposure to dilute buffers make substantial loss of proteins highly probable. We might add that if the nuclei are allowed to swell, the extrachromosomal network volume is presumably increased, which will not only accelerate the loss of 'soluble' components but also provoke the egress of molecules and complexes more or less weakly adsorbed to the chromatin or the nucleoskeleton. This might be relevant to the failure of swollen (as opposed to unswollen) nuclei to exhibit normal nuclear RNA restriction in isolation, a finding that has previously been recruited in support of the solid-state transport perspective.[64]

OVERVIEW

The theoretical demonstration that diffusion theory (however modified) cannot explain macromolecule transport in cells has a wealth of experimental corroboration, but it is nevertheless initially hard to accept. We have suggested some reasons for its entrenchment: the quantitative connotations of everyday language use of 'diffuse', the encouragement this gives to recurrent logical fallacies, the firm roots of diffusion theory in Maxwell-Boltzmann statistics and statistical mechanics. It is nevertheless seriously misleading when applied to cellular macromolecules, and fundamentally different models are needed to account for RNA and protein transport. We have outlined some possible models, and note that almost all of them involve the cytoskeleton or nucleoskeleton as either: (a) actively-participating transport systems; (b) direct or indirect organizers of heterogeneous assemblies of binding sites or (c) generators of net fluid movement. Further evaluation of these models is needed if we are to establish 'the meaning of nucleocytoplasmic transport', and the obvious next step is therefore to explore the structural aspects of cytoskeleton, nucleoskeleton, and other structures—notably the pore-complex—that might relate to macromolecule transport in cells. To take this step, we have to return to yet another problem of language-use that we have mentioned previously; what do we mean when we describe a structure as well or ill characterized?

REFERENCES

1. Fick AE. Über diffusion. Ann Phys Leipzig 1855; 94:59-86.
2. Einstein A. Von der molekulärkinetischen Theorie der Wärme

gefordete Bewegung von in ruhenden Flüssigkeiten suspendierten Teilchen. Ann Phys 1905; 17:549-554.

3. Smoluchowski M von. Zür Kinetischen Theorie der Brownschen Molekulärbewegung und der Suspensionen. Ann Phys 1906; 21:756-780.

4. Fourier JB. Théorie Analytique de la Chaleur. Paris: Oevres 1822. In: Translated Freeman A. The Analytical Theory of Heat. London: Cambridge University Press, 1878.

5. Carslaw HS, Jaeger JC. Conduction of Heat in Solids. Oxford: Clarendon, 1959.

6. Agutter PS, Malone PC, Wheatley DN. Intracellular transport mechanisms; a critique of diffusion theory. J Theoret Biol 1995; 176:261-272.

7. Hille B. Ionic Channels of Excitable Membranes. Sunderland, Massachusetts: Sinauer, 1984.

8. Donnan FG. Concerning the applicability of thermodynamics to the phenomena of life. J Gen Physiol 1927; 8:885-694.

9. Kellermeyer M, Rouse D, Gyorkey F et al. Potassium retention in membraneless thymus nuclei. Physiol Chem Phys Med NMR 1984; 16:503-511.

10. Wheatley DN. On the possible importance of an intracellular circulation. Life Sci 1985; 36:299-307.

11. Wheatley DN, Malone PC. Heat conductance, diffusion theory and intracellular metabolic regulation. Biol Cell 1993; 79:1-5.

12. Halling PJ. Do the laws of chemistry apply to living cells? Trends Biochem Sci 1989; 14:317-318.

13. Saxton MJ. Lateral diffusion in an archipelago. Biophys J 1982; 39:165-173.

14. Bers DM, Peskoff A. Diffusion around a cardiac calcium channel and the role of surface bound calcium. Biophys J 1991; 59:703-721.

15. Eckstein EC, Belgacem F. Model of platelet transport in flowing blood with drift and diffusion terms. Biophys J 1991; 60:53-69.

16. Zwanzig R, Szabo A, Bagchi B. Levinthal's paradox. Biophys J 1991; 60:671-678.

17. Fulton AB. How crowded is the cytoplasm? Cell 1982; 30:345-347.

18. Wolosewick JJ, Porter KR. Microtrabecular lattice of the cytoplasmic ground substance. J Cell Biol 1976; 82:114-139.

19. Srere P. The Metabolon. Trends Biochem Sci 1985; 10:109-111.

20. Srivastava DK, Bernhard S. Metabolite transfer via enzyme-enzyme complexes. Science 1986; 234:1081-1083.

21. Weiner N. Cybernetics. New York: Wiley, 1948.

22. Wheatley DN, Redfern A, Johnson RPC. Heat induced disturbances of intracellular movement and the consistency of the aqueous cytoplasm in HeLa S3 cells; a laser-Doppler and proton NMR study. Physiol Chem Phys NMR 1991; 23:199-266.

23. Wheatley DN, Clegg JS. What determines the basal metabolic rate

of vertebrate cells in vivo? Biosystems 1994; 32:83-92.

24. Gerard RW. Unresting Cells. New York: Harper, 1940.

25. Tyrrell HJV. Diffusion and Heat Flow in Liquids. London: Butterworth, 1961.

26. Crank J. The Mathematics of Diffusion. 2nd ed. Oxford: Clarendon Press, 1975.

27. Anderson DK, Hall JR, Babb AL. Mutual diffusion in non-ideal binary liquid matrices. J Phys Chem 1958; 62:404-409.

28. Irani RR, Adamson AW. Transport processes in liquid systems 3. Thermodynamic complications in the testing of existing diffusional theories. J Phys Chem 1960; 64:199-204.

29. Robinson RA, Stokes RH. Electrolyte Solutions. London: Butterworth, 1965.

30. Rice SA, Fisch HL. Statistical theory of transport. Ann Rev Phys Chem 1962; 11:187-272.

31. Polanyi M. Life's irreducible structure. Science 1968; 160: 1308-1312.

32. Paine PL, Horowitz SB. The movement of material between nucleus and cytoplasm. In: Goldstein L, Prescott DM, eds. Cell Biology: A Comprehensive Treatise. New York: Academic Press, 1980:299-338.

33. Wang Y-L, Lanni F, McNeil PL et al. Mobility of cytoplasmic and membrane-associated actin in living cells. Proc Natl Acad Sci USA 1982; 79:4660-4664.

34. Jacobson K. Wojcieszyn J. The translational mobility of substances within the cytoplasmic matrix. Proc Natl Acad Sci USA 1984; 81:6747-6751.

35. Clegg JS. Properties and metabolism of the aqueous cytoplasm and its boundaries. Amer J Physiol 1984; 246: R133-151.

36. Peters R. Fluorescence microphotolysis to measure nucleocytoplasmic transport and intracellular mobility. Biochim Biophys Acta 1986; 864:305-359.

37. Lang I, Scholz M, Peters R. Molecular mobility and nucleocyto-plasmic flux in hepatoma cells. J Cell Biol 1986; 102:1183-1190.

38. Horowitz SB, Fenichel IR, Hoffman B et al. The intracellular transport and distribution of cysteamine phosphate derivatives. Biophys J 1970; 10:944-1010.

39. Horowitz SB. The permeability of the amphibian oocyte nucleus in situ. J Cell Biol 1972; 54:609.

40. Horowitz SB, Miller DS. The intracellular distribution of adenosine-triphosphate. In: Welch GR, Clegg JS, eds. The Organization of Cell Metabolism. NATO ARI Series A B127. New York: Plenum Press, 1987:79-85.

41. Paine PL. Diffusive and nondiffusive proteins in vivo. J Cell Biol 1984; 99:188s-195s.

42. Chambers R. The micromanipulation of living cells. In: Moulton FR, ed. The Cell and Protoplasm. AAAS Publication 14. Washing-

ton DC: 1940:49-67.

43. Feldherr CM, Ogburn JA. Mechanisms for the selection of nuclear polypeptides in *Xenopus* oocytes. II: Two-dimensional gel analysis. J Cell Biol 1980; 87:589-593.

44. Dreyer C, Wang Y-H, Wedlich D et al. Oocyte nuclear proteins in the development of *Xenopus*. In: Hausen P, McLaren A, eds. British Society for Developmental Biology Symposium. London: Cambridge University Press, 1983:285-331.

45. Findlay DR, Newmeyer DD, Price TM et al. Inhibition of in vitro nuclear transport by a lectin that binds to nuclear pores. J Cell Biol 1987; 104:189-200.

46. Adam SA, Lobl TJ, Mitchell MA et al. Identification of specific binding proteins for a nuclear location sequence. Nature 1989; 337:176-179.

47. Zasloff MA. tRNA transport from the nucleus in a eucaryotic cell: carrier-mediated process. Proc Natl Acad Sci USA 1983; 80: 6436-6440.

48. McDonald JR, Agutter PS. The relationship between polyribonucleotide binding and the phosphorylation and dephosphorylation of nuclear envelope protein. FEBS Lett 1980; 116:145-148.

49. Bernd A, Schröder H-C, Zahn RK et al. Modulation of the nuclear envelope NTPase by poly(A) rich mRNA and by microtubule proteins. Eur J Biochem 1982; 129:43-49.

50. Maul GG. The nuclear and cytoplasmic pore complex: structure, dynamics, distribution and evolution. Int Rev Cytol Suppl 1977; 6:75-186.

51. Wojcieszyn JW, Schlegel RA, Wu ES et al. Diffusion of injected macromolecules within the cytoplasm of living cells. Proc Natl Acad Sci USA 1981; 78:4407-4410.

52. Wojcieszyn JW, Schlegel RA, Jacobson KA. Measurements of the diffusion of macromolecules injected into the cytoplasm of living cells. Cold Spring Harbour Symposium on Quantitative Biology 1981; 46:39-44.

53. Luby-Phelps K. Physical properties of cytoplasm. Curr Opin Cell Biol 1994; 6:3-9.

54. Gershon N, Porter K. Trus B. The microtrabecular lattice and the cytoskeleton: their volume, surface area and the diffusion of molecules through it. In: Oplatka A, Balaban M, eds. Biological Structures and Coupled Flows. New York: Academic Press, 1983: 377-380.

55. Sheetz MP, Spudich JA. Movement of myosin-coated fluorescent beads on actin cables in vitro. Nature 1983; 303:31-35.

56. Remenyik CJ, Kellermeyer M. A fluid mechanical hypothesis for macromolecule transport in living cells. Physiol Chem Phys 1978; 10:107-113.

57. Coulson RA. Metabolic rate and the flow theory: a study in chemi-

cal engineering. Comp Biochem Physiol 1986; 84A:217-229.

58. Welch GR, Clegg JS, eds. The Organization of Cell Metabolism. NATO ARI Series A B127. New York: Plenum Press, 1987.

59. Maul GG, ed. The Nuclear Envelope and the Nuclear Matrix. New York: Alan R Liss, 1982.

60. Agutter PS. Models for solid-state transport: messenger RNA movement from nucleus to cytoplasm. Cell Biol Internat 1994; 18:849-858.

61. Zachar Z, Kramer J, Mims IP et al. Evidence for channeled diffusion of pre-mRNAs during nuclear RNA transport in metazoans. J Cell Biol 1993; 121:729-742.

62. Jackson DA, Hassan AB, Errington RJ et al. Replication and transcription sites are colocalized in human cells. EMBO J 1993; 12:1059-1065.

63. d'Angelo EG. Micrurgical studies on *Chironomous* salivary gland chromosomes. Bull Biol 1946; 90:71-87.

64. Agutter PS. Nucleocytoplasmic mRNA transport: a plea for methodological dualism. Trends Cell Biol 1994; 4:278-279.

CHAPTER 3

CELLULAR STRUCTURES AND NUCLEOCYTOPLASMIC TRANSPORT

INTRODUCTION

'Well-characterized' has no precise or stable agreed denotation; its semantic content, if any, derives from context. Generally it carries the connotation that good science has been done and progress has been made. Conversely, 'X is poorly-characterized' implies that our knowledge does not provide an adequate basis for claims that might be advanced about X; it usually heralds a skeptical assessment of someone else's work. Sometimes these phrases are used emotively, serving only laudatory or pejorative functions. So much for precision; the point about stability is even more obvious. In the 1950s an enzyme was well-characterized if it had been purified biochemically and its kinetic properties and inhibitor sensitivities had been quantified; nowadays, sequence data, three-dimensional structure, active site chemistry and regulation of expression seem to be minimal requirements and kinetic information is of secondary interest. A method is well-characterized if it yields results that are judged reproducible, valid and interpretable; but as new methods supersede old, presumably a once-accepted procedure becomes at least relatively ill-characterized. Fashions in scientific method come and go, and as we have already seen this makes it dangerously tempting to dismiss work based on 'outdated' methods. Perhaps it is worth recalling that modern astronomers do not disregard Kepler, Herschel or Hubble merely because techniques have improved since then.

We need to spell out exactly how we intend to use 'well-characterized' in this book if we are to deploy the phrase intelligibly. Its application to supramolecular structures is particularly problematic because the biochemical basis of morphology remains an infant field. The reasons for this merit a brief review. However, the problem cannot be avoided because it is clear from chapters 1 and 2 that supramolecular structures play direct or indirect parts in nucleocytoplasmic transport. We are bound to ask what part each structure plays and how it acts and however we choose to phrase these questions they evoke the issue of 'characterization.'

MORPHOLOGY AND BIOCHEMISTRY

Those who were undergraduates as recently as the late 1960s can recall the mutual suspicion then obtaining between morphologists and biochemists; each accused the other of dealing in artifacts and declined to take the rival camp's discoveries seriously, a barrier which Needham had lamented in the 1930s.[1] The situation changed around 1970 when certain biological entities, notably elements of the cytoskeleton, came to be regarded as identical in centrifuge tube and electron microscope specimen.[2,3] Since then the biochemical basis of supramolecular structure up to the level of tissue organization has become an increasingly serious area of inquiry.

This historical attitude change has largely resulted from technical advances, which always play important roles in the direction of research. For example, almost all the techniques available to biochemists in the first half of the 20th century were those of the classical organic chemist, so this period saw the identification of the chemical structures of vitamins and other low M_r biomolecules and the elucidation of metabolic pathways. Then the emergence of polymer chemistry and biophysical techniques such as X-ray diffraction of macromolecules enabled protein and nucleic acid structures to be clarified, and this was a necessary condition for the emergence of molecular biology. (It was not a sufficient condition; the discoveries of classical and microbial genetics were equally important, and so was the borrowing of conceptual and linguistic metaphors from the then nascent field of control engineering). Later, techniques such as immunofluorescence and immunoelectron microscopy, together with advances in protein gel electrophoresis, opened the door to the biochemical study of supramolecular structures. More recently, gene manipulation methods and monoclonal antibodies have accelerated progress in this field and provided essential complements to morphological investigations.[3]

The direct study of ultrastructure has undergone parallel transformations. Traditional histochemistry dominated in the first half of the 20th century and enabled morphological features to be categorized only in terms of differential dye affinities and such shapes as could be resolved by light microscopes.[4] The advent of the electron microscope vastly extended the limits of resolution and since the middle of the century the development of new electron microscopic techniques has repeatedly heralded advances in knowledge. Salient examples are increasingly specific staining procedures such as immunogold, the introduction of scanning and high energy transmission microscopes, probe microanalysis and stereoimaging.[5-7] Computer image analysis of biopolymers is being applied with increasing frequency and success.[8] Thus, the ultrastructures of specimens of exactly known chemistry are being investigated along with the chemistries of materials of exactly known ultrastructure.

In so far as 'biochemical morphology' can be said to label a distinctive field of contemporary biology, it is a wide and heterogeneous field, perhaps dominated by research into mammalian extracellular matrices. In this book we are concerned almost exclusively with intracellular aspects of morphology, but we should address the topic according to the conceptual orientation of the whole field. This orientation seems to be reducible to three general principles. First, the organization of any structure is determined by specific bindings amongst its macromolecular constituents. Second, because these bindings depend mainly on short-range noncovalent forces, their frequencies and orientations are determined by the shapes of the individual macromolecules (the spatial distributions and accessibilities of binding sites therein). Third, assembly and disassembly of a structure are regulated by changing the availabilities and accessibilities of one or more of these sites (e.g., by sequestering components or phosphorylating key parts of a polypeptide sequence). Understanding the biochemical basis of morphology therefore entails knowing the three-dimensional shapes of structural macromolecules, the locations and specificities of their interactive sites, and the control of accessibility/availability of these sites.

WHEN IS A STRUCTURE WELL-CHARACTERIZED?

This preamble enables us to consider more critically what we actually assert when we describe a supramolecular structure as 'well-characterized'. The first criterion is obviously the recognition of a reproducible in situ ultrastructure. If there is no definite structure there is no more to be said. Under defined preparation conditions the electron microscope should reveal identical structures in comparable specimens; preferably, several different preparation procedures should give compatible images. The structures are judged identical in terms of such parameters as fibril diameter and the regular spacing and size of granules or other features. Obviously this criterion is purely morphological. Interactions with specific electron-opaque reporters such as gold-labeled antibodies might be used to establish identity or reproducibility, but logically this type of procedure evokes a separate and more complex criterion (see below).

Second, at least for historical completeness, we should include the recognition of an in vitro preparation identical with the structure visualized in situ. This second criterion was integral to the achievement of 'well-characterized' status for cytoskeletal elements, but may have lost its force during the past two decades. For example, the purification by biochemical tissue-fractionation methods of a preparation morphologically indistinguishable from in situ microtubules was a landmark achievement.[2] In contrast, much laboratory time has been expended in attempts to purify pore-complexes,[9] but this has now ceased to be a major target of research, although a fraction from yeasts that is highly enriched in pore-complexes will doubtless prove valuable.[10] The change in emphasis here becomes comprehensible when we consider why the second criterion was previously regarded as important. Around 1970, once microtubules were characterized, gel electrophoresis sufficed to identify their major structural components, the tubulins; at that time in history preliminary purification of microtubules was an indispensable means to that end. Nowadays, with the possibility of probing permeabilized cells with reporter molecules highly specific for individual proteins, structural components can be identified without recourse to bulk in vitro procedures. It is this third criterion, identification of the main structural (mainly protein) components of a supramolecular entity, that is fundamental to a *biochemical* morphology. How the criterion is satisfied is of relatively minor importance so long as a scientific consensus accepts the conclusion. The modern in situ approach has the advantages of directness and material economy, though claims for greater 'naturalness' based on the relative maintenance of cellular integrity are dubious in view of the disturbance of the intracellular milieu by permeabilization and the introduction of large numbers of alien reporter molecules (see chapter 2). Its disadvantages are the difficulties of quantification (how do we know whether the identified component is major?) and kinetic analysis. Whatever the scientific pros and cons, the in situ approach has a distinct aesthetic advantage and it is fashionable, which is another way of saying that it affords the best chance of generating consensus at the present time.

However they are established, the first and third criteria jointly constitute the foundation for any further study. Once a supramolecular structure has been described and its main components have been identified, a definite research program can be delineated, guided by the general principles of biochemical morphology outlined in the previous section. The criteria to be satisfied now are establishment of the shape of each component, location of the sites of interaction between one component and another, consideration of possible modes of regulation at these sites (e.g., phosphorylation), and analysis of the stages of structural assembly and disassembly. Criterion number four, establishment of component shape, was traditionally achieved by applying protein chemical and biophysical techniques to material purified in bulk (and, ideally, crystallized). Nowadays, research in this area is dominated by gene isolation and cloning techniques supplemented by increasingly powerful structure prediction methods. It is worth emphasizing that this change from 'tradition' has been a phenomenon of the last two decades. Criterion number five, location of binding sites, was traditionally achieved by chemical modification of particular amino acid residues or tightly-bound ligands in the wild type protein but now relies heavily on specific gene modifications; measurements of binding with the chemically or genetically altered protein remains an indispensable step and is properly supported by ultrastructural evidence for or against interaction.

The other criteria have to do with the dynamics of the structure. Briefly, the sixth (site regulation) relies on biochemical methods such as ^{32}P incorporation and the identification of appropriate consensus sequences through databases. The seventh (assembly/disassembly) requires ultrastructural studies of in vitro mixtures of purified components or portions thereof. We might add an eighth criterion, referring to function; if a structure serves to anchor or transport some other intracellular object, the anchoring sites, motors etc. involved must be identified and located. These issues will concern us in more detail in chapters 4-5, but for the present we are concerned explicitly with structure and therefore mainly with criteria 1-5. That is to say, we shall consider a structure well-characterized if we have: (a) a reproducible and detailed ultrastructural description; (b) a list of major components; (c) a knowledge of the three-dimensional shape of each component and (d) a knowledge of the locations, specificities and affinities of sites binding one component to another. Logically, we cannot usefully address questions of dynamics until these criteria are satisfied; but in practice, functional and dynamic studies conducted in advance of complete characterization often generate hypotheses that elucidate structure. Science progresses through pursuing the methodologically possible rather than paying exclusive attention to the logically prior.

We argued in chapters 1 and 2 that the nucleoskeleton, pore-complex and cytoskeleton are directly or indirectly involved in nucleocytoplasmic transport. We wish to know how they are involved, i.e., what dynamic roles they play. The first prerequisite is to establish how well-characterized these structures are. Unless our answers are positive our position in regard to understanding the relevant dynamics is a fortiori weak.

CYTOSKELETAL ELEMENTS

Characterization of the various elements of the cytoskeleton was already well advanced in the 1970s and our current knowledge of them is certainly more complete than that of pore-complex and nucleoskeleton. We emphasize that we are leaving aside the question of actual involvement in nucleocytoplasmic transport for the present and addressing only the issue of structural characterization. We shall do this briefly. Microfilaments comprise crosslinked F-actin fibrils.[11] The 37 nm pitch helical array of monomers in the in situ fibril is reproduced in vitro and is more or less predictable from the tertiary structure of G-actin, and molecular details of interactions with actin-binding proteins to generate networks[12] with quasi-thixotropic mechanical properties[13] are fairly well understood. Moreover, the assembly/disassembly mechanisms of F-actin[14,15] and microfilament networks[16] are known in considerable detail; we might add that the coupling mechanisms between actin and its characteristic motors, the myosins, have been fully analyzed as well.[17] Therefore, the criteria for structural characterization have been almost wholly met (with reservations about interaction-site details for some actin-binding proteins) and aspects of dynamics have also been established. Questions remain about the regulation of assembly/disassembly mechanisms and network stability, though the roles of profilin and the calcium-gelosolin system are well established,[18] and the list of actin-binding proteins remains open-ended. In so far as actin cables serve as transport routes for bodies such as organelles, myosin-I isoforms (minimyosins) act as motors[19] though the exact molecular details of the coupling between motor and transport substrate have not been elucidated.

Characterization of microtubules is of similar status; that of intermediate filaments is less complete because both dynamic mechanisms and distinctive biological function remain elusive; but in terms of the criteria that most concern us in this chapter, here the picture is fairly full. The components of the nuclear lamina, the lamins, resemble cytokeratins closely and this structure is in some respects better-characterized than the intermediate filament system itself. For instance, it is established that assembly and disassembly depend crucially on the actions of site-specific lamin kinases and phosphohydrolases.[20,21] In somatic cells, the monomeric lamin B links the lamin A-C network to the inner nuclear membrane[22] but details of the evidently tight association between lamina and pore-complex remain obscure.[9] The intermediate filament and lamina systems might have high structural stabilities[24] because their components bind cooperatively.[21]

PORE-COMPLEXES

The ultrastructure of the pore-complex has been known in outline since the 1950s but recent advances in electron microscopy have revealed much more detail.[25-27] Current reviews suggest that there is now a reasonable consensus about the morphology, as represented schematically in Fig. 3.1,[28,29] and it is agreed that the ultrastructure is highly conserved amongst eukaryotes.[30] The core of the assembly is a central 'plug' which may be a tubule of variable patent diameter. A ring (the inner spoke ring) surrounds the waist of this 'plug' and is buttressed to struts that connect it to the pore margin, probably to a pair of rings that surround the cytoplasmic and nuclear peripheries of the structure. These rings in turn are linked to the outer spoke ring, which lies within the perinuclear cisterna, the lumen enclosed by the fused nuclear mem-

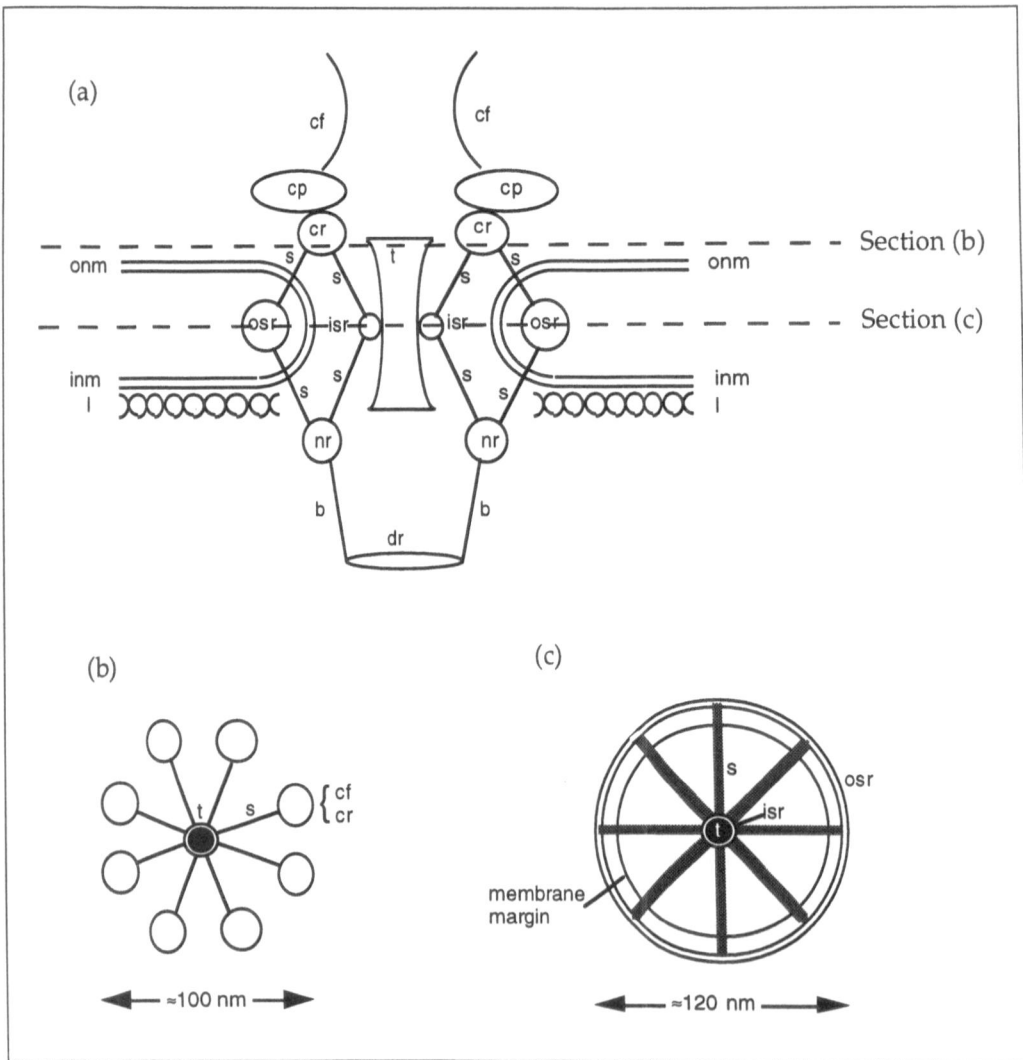

Fig. 3.1. Schematic representations of pore-complex. (a) Longitudinal section. (b, c) Radial sections (planes indicated in (a). Key: cf = cytoplasmic fibril; cp = cytoplasmic granule (particle); cr = cytoplasmic ring; t = central plug (tubule); s = spoke (strut); osr = outer spoke ring; isr = inner spoke ring; nr = nuclear ring; b = basket; dr = distal ring; onm = outer nuclear membrane; inm = inner nuclear membrane; l = lamina.

branes. Fibrils project outwards from both the nuclear and cyto-plasmic rings. The cytoplasmic fibrils may originate from granules associated with the ring; the nucleoplasmic ones form a basket structure which appears to terminate in a distal ring. This morphological complexity implies the involvement of many different structural components, though the overall octagonal symmetry and the repetitions of certain elements might suggest that a few are present in relatively high abundance. These inferences are generally supported by studies on *Saccharamyces cerevisiae* and vertebrates (human, rat, mouse and *Xenopus*), which are revealing an increasing number of pore-complex polypeptides of which two in yeast seem especially abundant. The compositional complexity is an obvious barrier to characterization.

The first pore-complex protein to be identified was the abundant high M_r concanavalin-A reactive glycoprotein gp210, which is located on the pore margin, largely inside the perinuclear cisterna.[31] Subsequently a series of *nucleoporins* were discovered by virtue of their N-acetylglucosamine containing O-linked carbohydrate moieties that make them reactive to wheat germ agglutinin.[32] The list of pore-complex proteins continues to grow. Evidence for the pore-complex location of a newly-identified polypeptide might include immunogold labeling, cofractionation with pore-complex enriched fraction, punctate immunofluorescent staining of the nuclear periphery, sequence homology with established components, or coordination of gene expression with other components. Of these approaches, immunoelectron microscopy can lead to more precise location within the structure. Sequence homology is useful because homologies with proteins located in other parts of the cell are very limited, and many of the pore-complex proteins identified to date share at least one of three repeat motifs. These motifs are GLFG, xFxFG (where x is any residue) and FG. GLFG-rich regions tend to be uncharged and N + Q rich. Regions with abundant xFxFG repeats are more highly charged and relatively poor in N and Q. FG-dominated sequences are poor in both charged and amide residues. S and T are abundant in all these repeat-motif regions, providing O-links for N-acetylglucosamine. Tables 3.2 and 3.3 summarize the structural information available at the time of writing.

Despite the considerable evolutionary conservation of pore-complex polypeptides there are uncertainties about comparisons between yeast and vertebrate components. For example, yeast NSP-1

Table 3.1. Characterization of cytoskeleton

Characteristic	Microfilaments	Microtubules	Intermediate Filaments
Ultrastructure	Cross-linked network or parallel arrays of 5-6nm double-helical filaments, 37nm pitch	Bundles of 50nm hollow helical fibers with multiple ends	Dense (mainly perinuclear) mesh of 7-8nm fibrils
Main components	G-actin. Cross-links in network involve e.g., filamin; anchorage in membrane by tolin/vinculin etc.	α and β tubulin (γ tubulin stabilizes cylinders) Microtubule-associated proteins and tau interlink bundles (details incomplete)	Cytokeratins, vimentin, desmin, neurofilamins or glial acidic protein (depending on tissue type) Ancillary proteins obscure; desmoplakin and related polypeptides act as desmosomal anchors
Site-site interaction	Mainly known	Mainly known	Mainly known
Assembly-disassembly mechanism	ATP-dependent treadmilling	Dynamic instability	?
Assembly-disassembly regulation	Actin phosphorylation Calcium-gelsolin mechanism Profilin sequestration of G-actin	Tubulin phosphorylation	Cytokeratin phosphorylation?
Associated motors	Myosin II (contractile) Myosins I (transport)	Dyneins (motility) Kinesins (transport)	None?

Table 3.2. *Pore-complex polypeptides with repeat motifs*

Polypeptide	Source	$M_r \times 10^{-3}$	Residue Span	Number of Repeats of Motif			Percentage Other Residues in Span				pI
				GLFG	xFxFG	FG	D+E	K+R	N+Q	S+T	
NUP 49	Yeast	49	1-267	13	1	3	0	3	24	23	5.9
NUP 100	Yeast	100	1-610	30	1	14	0	2	28	27	9.4
NUP 116	Yeast	116	166-748	33	0	9	0	2	29	20	9.3
			1-165	0	1	9	4	4	18	25	
NUP 145	Yeast	145	1-217	12	1	0	0	3	21	30	5.5
NSP 1	Yeast	86	1-175	4	1	8	0	2	27	32	6.1
			176-603	0	19	3	15	16	7	21	
NUP 1	Yeast	113	331-949	11	24	1	9	11	10	28	10.2
			950-1076	1	0	4	1	6	20	23	
NUP 2	Yeast	78	181-568	0	16	0	13	13	11	28	7.1
NUP 159[a]	Yeast	159	1-1460	0	3	25	15	10	10	21	4.7
p62[b]	Vertebrate	55	1-350	0-2	7-9	1-4	0	1	4-7	35-40	5.1
NUP 153	Vertebrate	153	880-1468	0	21	10	5	7	10	31	8.9
POM 121	Vertebrate	121	801-1199	2	7	14	1	2	6	34	10.2
NUP 214	Vertebrate	214	1632-2090	3	6	31	1	2	10	32	7.2
NUP 358[a]	Vertebrate	358	1-3224	0	12	10	14	12	9	16	6.1

a: predicted from gene sequence
b: some species differences

Table 3.3. Other vertebrate pore-complex polypeptides

Polypeptide	pI	Repeat Motif	Glycosylation	Phosphorylation Sites	Comments
p54 p58	6.7 >8	All three may be present	Multiple O-linked	None?	Form stable complex with p62
gp210	6.5	None	N-linked	None?	Major component. No homology with other components
p260	5.0	None	None	?	Acidic C-terminal domain. Cytoplasmic orientation
p180	6.2	None	None	?	Cytoplasmic orientation
NUP155	6.0	?	?	Multiple	Unique sequence. Located on both faces
NUP107	5.4	?	?	None?	Unique sequence. C-terminal leucine zipper
NUP153	8.9	xFxFG, C-term	Multiple	Multiple	Four zinc fingers, C_2C_2 type. Basket locations

is homologous throughout its length with vertebrate p62[33,34] but functional similarity has not been established though both p62 and NSP-1 bind to other nucleoporins through an α-helical coiled rod domain.[35] These uncertainties are frustrating because *S. cerevisiae* is such a valuable experimental model; much of its genome has been sequenced, numerous mutants can be obtained almost to order, and yeast biochemistry and cell biology are highly developed fields, so it is likely that the biochemical morphology of the pore-complex will mature first in studies of this organism. If we provisionally assume the yeast-vertebrate homologies that are apparent (and, more safely, homologies between mammals and amphibia), then the emerging picture of the molecular architecture is more or less as shown in Fig. 3.2. Immunogold electron microscopy of sectioned cells colocates the pore margin proteins gp210, POM-121 and POM-152 in the outer spoke ring within the perinuclear cisterna[36,37] and this ring may be made of a complex of these three proteins dominated by gp210, which is present in some 25 copies per pore-complex.[31] The same technique locates the nuclear pore proteins NUP-180, NUP-214 and p260 in the cytoplasmic fibrils of subcellular fractions,[38,39] NUP-358 (vertebrate)[40] and NUP-159

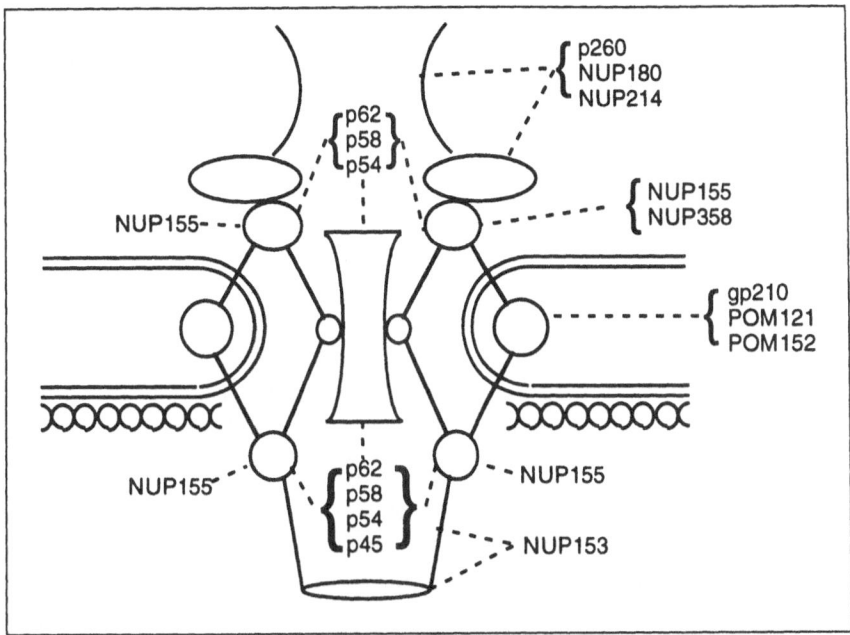

Fig. 3.2. Possible locations of some pore-complex polypeptides. The pore margin (POM) polypeptides and the abundant gp210 are probably outer spoke ring components. The nuclear pore protein(NUP) 153, which has C_2C_2 zinc fingers, is located in the nuclear basket or distal ring and NUP-358 in the cytoplasmic ring. Since p62 occurs on both pore faces, in either the nuclear and cytoplasmic rings or the central plug, it is likely that p58, p45 and p54 are similarly located; they form a stable complex with p62. NUP-180, NUP-214 and p260 are probably located in the cytoplasmic fibrils and/or granules. The location of NUP-155 is less clear but since it is exposed on both faces of the pore the nuclear and cytoplasmic rings are likely candidates.

(yeast)[41] in the cytoplasmic ring, p62 possibly on the cytoplasmic and nuclear rings or more probably on both faces of the central plug[42-46] and NUP-153 in the nucleoplasmic basket.[47] Because an apparently fibrillar complex of M_r 230,000 has been isolated comprising a stable complex of p62, p58, p45 and p54, these four proteins may be colocated in situ as shown in Fig. 3.2.[42] Certainly p62 is located in more than one position within the pore-complex because it has been extracted from both the cytoplasmic and nucleoplasmic faces.[42-45] Another protein with this Janus distribution is the distinctive NUP-155, not homologous to other polypeptides and endowed with multiple phosphorylation sites.[48] In Figure 3.2 we tentatively locate it in the nuclear and cytoplasmic rings.

Sequence data are broadly consistent with these assignments. POM-152, POM-121 and gp210 each contain hydrophobic stretches which are probably membrane spanning regions, and there is biochemical evidence that they are integral membrane proteins.[31,37,49] These features suggest that the luminal complex might fix the orientation of the pore structure within the membrane, particularly if the cytoplasmic domain of POM-121 binds, for example, the cytoplasmic granules or ring (there is no evidence for this possibility but none to exclude it either). NUP-153 contains four C_2C_2-type zinc finger motifs and binds DNA in vitro.[47,50] Its spatial location affords possibilities for nucleic acid binding in vivo. More generally, most of the protein sequences contain typically filamentous motifs such as those characteristic of helical coiled-coils; indeed the p62-p58-p54-p45 complex has been vizualised as a 50 nm fibril with 15 nm terminal rings of p62.[42,43] Since the rings, filaments and struts of the pore-complex are fibrillar, this general observation together with specific location data suggests that a detailed biochemical morphology is emerging.

However, it is far from complete as yet. Several proteins remain to be located even provisionally, their specific interactions and even their binding stoichiometrics remain largely unknown, and it is not clear whether homologous motif-repeat regions interact among different polypeptides to form coherent domains. There are no real clues about the mechanisms of binding of the pore-complex to the lamina or other structures. Finally, the role of the carbohydrate moieties of the glycoproteins is unclear. Alteration of the carbohydrate has no identifiable effect on transport functions or structural integrity.[51] Perhaps the extended highly hydrated sugar-rich regions make the overall structure rigid enough to withstand deformations during cell movement that might impair function.

Functional studies might supplement this structural knowledge, a possibility that we shall pursue in chapter 4. The aspect of dynamics most relevant to basic characterization of any structure concerns assembly-disassembly mechanisms and these are certainly pertinent to our understanding of pore-complex biology. The structures disassemble and reassemble during open mitosis and their density on the nuclear surface is maintained during interphase and throughout the yeast cell-cycle by de novo synthesis or elimination.[52,53] The mechanisms are largely undefined but the literature contains at least three suggestive pieces of information. One pore-complex protein (NUP-155) is known to contain multiple phosphorylation sites[48] and others may do likewise, and it is tempting to suppose that disassembly might be phosphorylation-dependent as in the lamina (see above). Another component whose location is unknown, NUP-107, contains an apparent C-terminal leucine-zipper sequence and may therefore have a role in assembly.[54] Finally, disorganized components stockpiled in egg cytoplasm can be induced to form nuclear structures including functional pore-complexes by the addition of random naked DNA.[55]

Comparison of this survey with Table 3.1 reveals the contrast between cytoskeletal elements and pore-complexes in respect of structural characterization. The ultrastructure is fairly well established; some components have been identified, though whether these include 'main' ones is not certain; site-site interaction details are at best partly known; and so are the mechanisms and regulation of assembly and disassembly. Granted that noone seriously doubts that pore-complexes have a role in nucleocytoplasmic transport, this is not a happy situation.

NUCLEOSKELETON

Several terms have been invented to denote a presumed extra-chromosomal fibrillar framework in the nucleoplasm. We find it convenient to reserve the word 'nucleoskeleton' for the putative structure in vivo and the phrase 'nuclear matrix' for in vitro preparations; using different terms helps to guard against a possibly misleading assumption of identity.

The status of the nucleoskeleton remains uncertain, though 'controversy' would be an unsuitable word for a matter about which there is little active contemporary debate. The balance of faith appears to have tilted away from the predominant skepticism of the 1980s, but many workers still (for example) address the questions of supranucleosomal chromatin organization without reference to a nucleoskeleton, anticipating that an appropriate convergence of molecular and ultrastructural data will generate models lacking any extrachromosomal framework. Moreover, references to the nucleoskeleton in discussions of nucleocytoplasmic transport are still quite rare. For example, much current literature shows a sequence of interesting events, each meriting elucidation, occurring at the pore-complex during karyophilic protein import (chapter 4), but implies by default that once the protein is out of the pore-complex again it finds its specific intranuclear target spontaneously and with no further ado. This view ignores the fact that the target might be several microns away in a crowded and complex environment, and also overlooks the now widely-accepted association between the basket structure of the pore-complex and an underlying nuclear framework.[30,47,50] There are probably two related reasons for this seeming neglect: the nucleoskeleton, if it exists, remains poorly-characterized; and experiments designed to probe intranuclear events involving such a structure resist unequivocal interpretation. On the other hand, the exclusive devotion of a recent volume of *Int Rev Cytol* to the nucleoskeleton is symptomatic of changing attitudes, suggesting that the structure is at least en route to characterization.

The first criterion, an ultrastructural description in situ, has probably been satisfied but no single piece of evidence is unchallengable. High-energy transmission electron micrographs of thick resinless sections[56,57] indicate a reticular network of approximately 10 nm fibrils (with some finer ones) extending throughout the nucleus, but such images are notoriously difficult to interpret despite their aesthetic qualities. Compatible images have been obtained from conventional thin sections but only after constriction of the chromatin with α-amanitin,[58] which could generate artifacts. Immunoelectron microscopic localization of the high M_r HIB2 antigen,[59] the 68,000 molecular mass phosphoprotein B2[60] and other polypeptides[61] to an in situ nucleoskeleton could conceivably be interpreted by reference to high-order (residual) chromatin structure. The second criterion, isolation of a morphologically identifiable subcellular fraction, might have been met by gradual and progressive extraction of chromatin from nuclei under conditions controlling the artifactual crosslinking of proteins,[62,63] but early nuclear matrix preparations[64] were not morphologically well formed. Because of this, it was widely held for many years that the nuclear matrix was an artifact of preparation[65] or at best an operationally defined fraction whose isolation implied nothing definite about nuclear organization in vivo. It certainly seemed difficult to explain why no nuclear matrix was observed when nuclei were disrupted with heparin[66] or by DNAase I[67] or II[68] at low ionic strengths, and why no connecting framework was seen in Miller spreads,[69] and evidence that sulfhydryl group oxidation in tissue homogenates was necessary to stabilize the nuclear matrix[70] supported the view that matrix preparations were artifacts.

The recent decrease of skepticism about the reality of the nucleoskeleton seems to have been occasioned partly by evidence for association between the pore-complex basket and an intranuclear lattice,[30,45] particularly the observation that NUP-153 is located in intranuclear fibrils,[50] and partly by the partial satisfaction of the third criterion of characterization. Short F-actin fibrils appear to be major nucleoskeletal components[71,72] and they are probably intimately associated with nuclear RNA. Certainly both RNAases and cytochalasins disrupt the more morphologically well formed matrix preparations,[63] cytochalasins specifically release splicing intermediates,[73,74] and specific associations have been reported between actin, snRNPs and the splicing factor SC-35.[75,76] In addition, cytochalasin B releases the HnRNP core C-group polypeptides from nuclear subfractions[74] and these proteins have been implicated in linking HnRNA to the nucleoskeleton.[77] If F-actin is indeed an integral part of the structure then the absence of any fibrils after exposure to DNAase I or low salt concentrations is easily explained. DNA topoisomerase II has also been proposed as an integral nucleoskeletal component,[78] but although this is consistent with the role of the structure in DNA anchoring and the initiation of replication and transcription[79,80] the evidence is stronger for *Drosophila* than for other organisms. Two other possible nucleoskeletal components, PI1 and PI2,[60] seem to be required for normal genomic expression.[81] Finally, the nuclear protein associated with the mitotic apparatus (NuMA) is believed to be a major nucleoskeletal component during interphase; it contributes to the establishment and maintenance of a bipolar spindle during mitosis.[82] NuMA is present in some 2×10^5 copies per cell and has an M_r of 240,000,[83,84] each molecule comprising globular head and tail domains separated by a 1500 Å α-helical rod containing largely hydrophobic heptad repeat motifs. Despite this cytokeratin-like rod sequence the interphase distribution of NuMA is shown by immunofluorescence to be transnuclear, not lamina-restricted.[83] The head and tail have no significant homologies with other polypeptides. These features (high abundance, transnuclear interphase distribution and a long domain likely to oligomerize into fibrils through coiled-coil interactions) suggest that NuMA may be a major nucleoskeletal component, and indeed immunogold staining locates it in many though not all the reticular fibrils visualized in resinless sections.[85] Moreover, there is powerful evidence that

NuMA is necessary for the formation of normal daughter nuclei at telophase.[86-88] Immuno-precipitation and immunofluorescence evidence for association between NuMA and splicing complexes[89] may be controversial,[90] but the protein has also been implicated in compartmentalizing EGF receptor messenger to the perinuclear region,[91] which suggests some association with nuclear RNAs. Whether this association is direct or via actin (see above) is unclear. It is possible that the apparently NuMA-free fibrils revealed by immunogold are made of F-actin[85] and that the cytochalasin and RNAase resistant fibrils in isolated matrix preparations[63] are made of NuMA, but this has yet to be investigated. Recently, evidence has been obtained for a transnuclear rather than an exclusively peripheral distribution of lamins throughout interphase,[92] and therefore lamins as well as NuMA could be structural components of the nucleoskeleton. Conversely, of the three known isoforms of NuMA, products of differential splicing of transcripts of a single gene assigned to human chromosome 11q13,[93] two appear to be cytoplasmic during interphase.[94]

Not surprisingly, nothing of any note can be said about the assembly-disassembly dynamics of the nucleoskeleton. NuMA requires both its head and tail domains to be intact before it can participate in normal nuclear assembly, but the head seems not be required for either interphase nuclear uptake or for association with spindle microtubules.[88] The possibility that the head binds directly or indirectly to F-actin merits investigation. Whether oligomerization is controlled by phosphorylation as for other intermediate filament-type proteins is unclear; a known phosphorylation site in the tail domain is necessary for spindle binding[95] but might not be relevant during interphase.

The nucleolus is a specialized region and its skeleton may be supplemented by distinctive components. These probably include polypeptides of M_r 38,000 (B23),[96,97] 145,000[98] and 180,000.[99] The nucleolar skeleton is continuous with the nucleoskeleton[100] and its periphery includes the dominant nucleolar structural proteins fibrillarin and nucleolin/C23.[101] Fibrillarin is a component of U2snRNP.[102] It is a basic protein of M_r 37,000 and is mainly restricted to the transition zone between the fibrillar center and the dense fibrillar component.[103,104] Nucleolin is another basic protein of 713 residues arranged into two structural regions; a chromatin-binding N-terminus and a C-terminus comprising four RNA-binding domains and a GR-rich sequence implicated in nucleoprotein packaging.[105] Along with B23 it accounts for much of the protein mass of the interphase nucleolus.[106] It seems to be implicated in both nuclear matrix attachment[107] and binding to the 3' untranslated region of pre-mRNAs[108] along with the C-group proteins of the ribonucleosome core.[77] It is ubiquitous amongst 'higher' eukaryotes and has a yeast homologue, NSR-1.[109]

CHARACTERIZATION AND TRANSPORT FUNCTION

Neither the pore-complex nor the nucleoskeleton is as well-characterized as the cytoskeleton, as summarized in Table 3.4. The significance of this for our overall theme will become increasingly apparent as we proceed, but a general point can be made here: either nucleocytoplasmic transport processes are solid-state, in which case the generation of movement by supramolecular structure has to be explained, or they are not, in which case the dynamics of these same structures underpin the relevant adsorption, exclusion and flow-generation events in macromolecule migration (chapter 2). In either case the characterization of intracellular organization is pertinent to our understanding, and since this characterization is incomplete our understanding must be incomplete as well. Even our knowledge of the microfilament system proves limited; the intracellular regulation of actin filament crosslinking remains a speculative issue[110-112] but one that is clearly relevant to cytoplasmic macromolecule movements, not least because of the extent of the microfilament system in most cell types.

The phrase 'solid-state transport' does not denote a unique type of mechanism. The commonest but perhaps least appropriate interpretation assumes an analogy with axonal transport (see chapter 1). If macromolecules are indeed exchanged between nucleus and cytoplasm by such a motor-driven mechanism then the motors and their attachments to the substrates need to be identified, but as we observed earlier the concept of a continuous nucleo-cytoplasmic fibril system is controversial.[50,57,113-117] However, cellular skeletons and their assembly/disassembly mechanisms are likely to be pertinent to macromolecule transport irrespective of whether the processes can properly be considered 'solid-state' (chapter 2). Therefore, until the characterization of supramolecular structures has advanced to the point where the mechanisms and regulation of assembly and disassembly are well established, intracellular

Table 3.4. Characterization of intracellular structures

Characteristic	Actin-Cytoskeleton	Nuclear Pore-Complex	Nucleoskeleton
Ultrastructure	Cross-linked network or parallel arrays of 5-6 nm double-helical filaments, 37 nm pitch	General consensus view: octagonally symmetric array of spokes linking inner spoke ring around central plug, outer spoke ring in pore margin, and nuclear and cytoplasmic rings; cytoplasmic fibrils and nuclear basket project	No settled consensus; probably an anastomosing meshwork of approximately 10 nm fibrils (some finer)
Main components	G-actin	More than 12 components identified and characterized; probable locations established for 7-11 of these	Short F-actin and NuMA are probable. Other possibilities are PI1, PI2, and perhaps DNA topoisomerase II and splicing factor SC-35, HIB2 and B2
Site-site interaction	Mainly known	Little known. p62-p58-p54 fibrillar complex partly characterized; membrane associations of pore-margin proteins provisionally established	NuMA oligomers may form; otherwise entirely unknown
Assembly-disassembly mechanism	ATP-dependent treadmilling	Unknown	Unknown
Assembly-disassembly regulation	Actin phosphorylation Calcium-gelsolin mechanism Profilin sequestration of G-actin	Unknown	Unknown
Associate motors	Myosin II (contractile)	Myosin possible in Drosophila,[109] otherwise unknown	Myosin possible?[110] Unknown

macromolecule transport processes will not be adequately understood.

This gloomy view of the status quo should not be taken to indicate that our impression of growing knowledge in the field is chimeric, but rather that premature conclusions are likely to be simplistic and misleading. In chapter 4 we shall explore this point by reviewing our current state of understanding of events at the pore-complex, with emphasis on nuclear protein import. In chapter 5 we shall extend the inquiry to nucleoskeleton and cytoskeleton, and here our emphasis will be on mRNA and its precursors. Subsequently we shall consider the light that certain pathologies of nucleocytoplasmic transport might shed on our understanding.

REFERENCES

1. Needham J. Biochemistry and Morphology. Cambridge: Cambridge University Press, 1934.
2. Amos LA, Baker TS. The three-dimensional structure of tubulin protofilaments. Nature 1979; 279:607-612.
3. Amos LA. Structure of muscle filaments studied by electron microscopy. Ann Rev Biophys Chem 1985; 14:291-313.
4. Clark WE. The tissues of the body. 6th ed. Oxford: Clarendon Press, 1971.
5. Siegel B, ed. Physical aspects of electron microscopy and microbeam analysis. New York: Plenum, 1975.
6. Duke PJ, Michette AG, eds. Modern microscopes. Techniques and applications. London: Plenum, 1990.
7. Spence JCH. Experimental high-resolution electron microscopy. Oxford: Clarendon Press, 1980.
8. Amos LA, Klug A. Arrangement of subunits in flagellar microtubules. J Cell Sci 1974; 14:523-549.
9. Aaronson RP, Blobel G. Isolation of nuclear pore-complexes in association with a lamina. Proc Natl Acad Sci USA 1975; 72: 1007-1011.
10. Rout MP, Blobel G. Isolation of the yeast nuclear pore complex. J Cell Biol 1993; 123:771-783.
11. Cooke R. The mechanism of muscle contraction. CRC Int Rev Biochem 1986; 21:53-118.
12. Stossel TP. Non-muscle actin binding proteins. Ann Rev Cell Biol 1985; 1:353-402.
13. Sato M, Schwartz WH, Pollard TD. Dependence of the mechanical properties of actin: alpha-actinin gels on deformation rate. Nature 1987; 325:828-830.
14. Korn ED, Cartier M-F, Pantaloni D. Actin polymerization and ATP hydrolysis. Science 1987; 238:638-644.

15. Tilney LG, Bonder EM, DeRosier DJ. Actin filaments elongate from their membrane-associated ends. J Cell Biol 1981; 90:485-494.

16. Matsudaira P, Janmey P. Pieces in the actin-severing protein puzzle. Cell 1988; 54:139-140.

17. Pollard TD. The myosin crossbridge problem. Cell 1987; 48: 909-910.

18. Yin HL. Gelsolin: calcium and polyphosphoinositide-regulated actin modulating protein. Bioessays 1987; 7:176-179.

19. Adams RJ, Pollard TD. Propulsion of organelles isolated from *Acanthamoeba* along actin filaments by myosin-I. Nature 1986; 322:754-756.

20. Gerace L, Blobel G. The nuclear envelope lamina is reversibly depolymerized during mitosis. Cell 1980; 19:277-288.

21. Georgatos S, Stournaras C, Blobel G. Heterotypic and homotypic associations between the nuclear lamins: site-specificity and control by phosphorylation. Proc Natl Acad Sci USA 1988; 85:4325-4329.

22. Aebi U, Kohn J, Buhle L et al. The nuclear lamina is a meshwork of intermediate-type filaments. Nature 1986; 323:560-564.

23. Worman HJ, Yuan J, Blobel G et al. A lamin B receptor in the nuclear envelope. Proc Natl Acad Sci USA 1988; 85:8531-8534.

24. Franke WW. Nuclear lamins and cytoplasmic intermediate filament proteins: a growing multigene family. Cell 1987; 48:33-34.

25. Akey CW, Radermacher M. Architecture of the *Xenopus* nuclear pore complex revealed by three-dimensional cryo-electron microscopy. J Cell Biol 1993; 122:1-19.

26. Goldberg MW, Allen TD. The nuclear pore complex: Three-dimensional surface structure revealed by field emission, in-lens scanning electron microscopy, with underlying structure uncovered by proteolysis. J Cell Sci 1993; 106:261-274.

27. Ris H, Malecki M. High-resolution field emission scanning electron microscope imaging of internal cell structures after Epon extraction from sections: a new approach to correlative ultrastructural and immunocytochemical studies. J Struct Biol 1993; 111:148-157.

28. Rout MP, Wente SR. Pores for thought: nuclear pore complex proteins. Trends Cell Biol 1994; 4:357-363.

29. Panté N, Aebi U. Towards understanding the three-dimensional structure of the nuclear pore complex at the molecular level. Curr Opin Struct Biol 1994; 4:187-196.

30. Miller M, Park MK, Hanover JA. Nuclear pore complex: structure, function, and regulation. Physiol Rev 1991; 71:909-949.

31. Gerace L, Ottaviano Y, Kondor-Koch C. Identification of a major polypeptide of the nuclear pore-complex. J Cell Biol 1982; 95:826-837.

32. Snow CM, Senior A, Gerace L. Monoclonal antibodies identify a group of nuclear pore-complex glycoproteins. J Cell Biol 1987;

104:1143-1156.

33. Hurt EC. A novel nucleoskeletal-like protein located at the nuclear periphery is required for the life-cycle of *Saccharomyces cerevisiae*. EMBO J 1988; 7:4323-4334.

34. Starr CM, d'Onofrio M, Park MK et al. Primary sequence and heterologous expression of nuclear pore glycoprotein p62. J Cell Biol 1990; 110:1861-1871.

35. Buss F, Stewart M. Macromolecular interactions in the nucleoporin p62 complex of rat nuclear pores: binding of nucleoporin p54 to the rod domain of p62. J Cell Biol 1995; 128:251-261.

36. Hallberg E, Wozniak RW, Blobel G. An integral membrane protein of the pore membrane domain of the nuclear envelope contains a nucleoporin-like region. J Cell Biol 1993; 122:513-521.

37. Wozniak RW, Blobel G, Rout MP. POM-152 is an integral protein of the pore membrane domain of the yeast nuclear envelope. J Cell Biol 1994; 125:31-42.

38. Greber UF, Senior A, Gerace L. A major glycoprotein of the nuclear pore complex is a membrane-spanning polypeptide with a large lumenal domain and a small cytoplasmic tail. EMBO J 1990; 9:1495-1502.

39. Wilken N, Kossner U, Senecal JL et al. NUP-180, a novel nuclear pore complex protein localizing to the cytoplasmic ring and associated fibrils. J Cell Biol 1993; 123:1345-1254.

40. Wu J, Manyunis MJ, Kraemer D et al. NUP-358, a cytoplasmically exposed nucleoporin with peptide repeats, Ran-GTP binding sites, zinc fingers, a cyclophilin A homologous domain, and a leucine-rich region. J Biol Chem 1995; 270:14209-14213.

41. Kraemer DM, Strambio de Castillia C, Blobel G et al. The essential yeast nucleoporin NUP-159 is located on the cytoplasmic side of the nuclear pore-complex and serves in karyopherin-mediated binding of transport substrate. J Biol Chem 1995; 270:19017-19021.

42. Guan T, Muller S, Klier G et al. Structural analysis of the p62 complex, an assembly of O-linked glycoproteins that localizes near the central gated channel of the nuclear pore complex. Mol Biol Cell 1995; 6:1591-1603.

43. Panté N, Bastos R, McMorrow I et al. Interactions and three-dimensional localization of a group of nuclear pore complex proteins. J Cell Biol 1994; 126:603-617.

44. Davis LI, Blobel G. Identification and characterization of a nuclear pore-complex protein. Cell 1986; 45:699-709.

45. Dabauvalle MC, Loos K, Scheer U. Identification of a soluble precursor complex essential for nuclear pore assembly in vitro. Chromosoma 1990; 100:56-66.

46. Cordes V, Waizenegger I, Krohne G. Nuclear pore complex glycoprotein p62 of *Xenopus laevis* and mouse: cDNA cloning and iden-

tification of its glycosylated region. Eur J Cell Biol 1991; 55:31-47.

47. Sukegawa J, Blobel G. A nuclear pore-complex protein that contains zinc finger motifs, binds DNA, and faces the nucleoplasm. Cell 1993; 72:29-38.

48. Radu A, Blobel G, Wozniak RW. NUP-155 is a novel nuclear pore complex protein that contains neither repetitive sequence motifs nor reacts with WGA. J Cell Biol 1993; 121:1-10.

49. Wozniak RW, Blobel G. The single transmembrane segment of gp210 is sufficient for sorting to the pore membrane domain of the nuclear envelope. J Cell Biol 1992; 119:1441-1449.

50. Cordes VC, Reidenbach S, Kohler A et al. Intranuclear filaments containing a nuclear pore complex protein. J Cell Biol 1993; 123:1333-1344.

51. Miller MW, Hanover JA. Functional nuclear pores reconstituted with beta1-4 galactose-modified O-linked N-acetylglucosamine glycoproteins. J Biol Chem 1994; 269:9289-9297.

52. Maul G. The nuclear and cytoplasmic pore-complex: structure, dynamics, distribution and evolution. Int Rev Cytol 1977; Suppl 6:75-186.

53. Forbes DJ. Structure and function of the nuclear pore complex. Ann Rev Cell Biol 1992; 8:495-527.

54. Radu A, Blobel G, Wozniak RW. NUP-107 is a novel nuclear pore complex protein that contains a leucine zipper. J Biol Chem 1994; 269:17600-17605.

55. Forbes DJ, Kirschner MW, Newport JW. Spontaneous formation of nucleus-like structures around bacteriophage DNA microinjected into *Xenopus* eggs. Cell 1983; 34:13-33.

56. Capco DG, Penman S. Mitotic architecture of the cell: the filament networks of the nucleus and cytoplasm. J Cell Biol 1983; 96:896-906.

57. Carmo-Fonescu M, Cicadao AJ, David-Ferreira JF. Filamentous cross-bridges link intermediate filaments to the nuclear pore-complexes. Eur J Cell Biol 1987; 45:282-290.

58. Brasch KR. Fine structure and localisation of the nuclear matrix in situ. Exp Cell Res 1982; 140:161-172.

59. Nickerson JA, Penman S. Localization of nuclear matrix core filament proteins at interphase and mitosis. Cell Biol Internat Rep 1992; 16:811-826.

60. Chew EC, Cheng-Chew SB, Deharven E et al. Distribution of a novel nuclear protein in normal and regenerating liver cells. In Vivo 1992; 6:97-102.

61. Chaly N, Bladon T, Setterfield G et al. Changes in the distribution of nuclear matrix antigens during the mitotic cell cycle. J Cell Biol 1984; 99:661-671.

62. Fey FG, Krochmalnic G, Penman S. The nonchromatin substructures of the nucleus: the RNP-containing and RNP-depleted ma-

trixes analyzed by sequential fractionation and resinless electron microscopy. J Cell Biol 1976; 102:1654-1665.

63. Comerford SA, Agutter PS, McLennan AG. Nuclear matrices. In: MacGillivray AJ, Birnie DG, eds. Nuclear Structures: their isolation and characterization. London: Butterworth, 1986:1-13.

64. Berezney R, Coffey DS. Identification of a nuclear protein matrix. Biochem Biophys Res Commun 1976; 60:1410-1417.

65. Arenstorf HP, Conway GC, Wooley JC et al. Nuclear matrix-like filaments form through artifactual rearrangements of HnRNP particles. J Cell Biol 1984; 99:233a.

66. Bornens M, Courvalin JC. Isolation of nuclear envelopes with polyanions. J Cell Biol 1978; 76:191-206.

67. Kay RR, Fraser D, Johnston IR. A method for the rapid isolation of nuclear membranes from rat liver. Eur J Biochem 1972; 30:145-154.

68. Krachmarov C, Tasheva B, Markov D et al. Isolation and characterization of nuclear lamina from Ehlich ascites tumor cells. J Cell Biochem 1986; 30:351-356.

69. Beyer AL, Bouton AH, Miller OL. Correlation of hnRNP structure and nascent transcript cleavage. Cell 1981; 26:155-165.

70. Kaufmann SH, Coffey DS, Shaper JH. Considerations in the isolation of rat liver nuclear matrix, nuclear envelope and pore-complex lamina. Exp Cell Res 1981; 132:105-123.

71. Nakayasu H, Ueda K. Small nuclear RNP complex anchors on the actin filaments in bovine lymphocyte nuclear matrix. Cell Struct Funct 1984; 9:317-326.

72. Schindler M, Jiang L-W. Epidermal growth factor and insulin stimulate nuclear pore-mediated macromolecular transport in isolated rat liver nuclei. J Cell Biol 1987; 104:849-853.

73. Schröder H-C, Trölltsch D, Friese U et al. Mature mRNA is selectively released from the nuclear matrix by an ATP/dATP-dependent mechanism sensitive to topoisomerase inhibitors. J Biol Chem 1982; 262:8917-8925.

74. Schröder H-C, Trölltsch D, Wenger R et al. Cytochalasin B selectively releases ovalbumin mRNA precursors but not the mature ovalbumin mRNA from hen oviduct nuclear matrix. Eur J Biochem 1993; 167:239-245.

75. Sahlas DJ, Milankov K, Park PC et al. Distribution of snRNPs, splicing factor SC-35 and actin in interphase nuclei: Immunocytochemical evidence for differential distribution during changes in functional states. J Cell Sci 1993; 105:347-357.

76. Carter KC, Bowman D, Carrington W et al. A three-dimensional view of precursor messenger RNA metabolism within the mammalian nucleus. Science 1993; 259:1330-1335.

77. Van Eekelen CAG, van Venrooij WJ. HnRNA and its attachment to a nuclear matrix. J Cell Biol 1981; 88:554-563.

78. Berrios M, Osterhoff N, Fisher PA. In situ localization of DNA topoisomerase II, a major polypeptide component of the *Drosophila* nuclear matrix. Proc Natl Acad Sci USA 1985; 82:4142-4146.

79. Berezney R. Dynamics of the nuclear protein matrix. In: Busch H, ed. The Cell Nucleus. Vol 7. New York and London: Academic Press, 1979:413-455.

80. Jackson DA, Cook PR. Transcription occurs at a nucleoskeleton. EMBO J 1985; 4:919-926.

81. Prather RS, Schatten G. Construction of the nuclear matrix at the transition from maternal to zygotic control of development in the mouse: An immunocytochemical study. Molec Reprod Devel 1992; 32:203-208.

82. Lyderson BK, Pettijohn DE. Human-specific nuclear protein that associates with the polar region of the mitotic apparatus: Distribution in a human/hamster hybrid cell. Cell 1980; 22:489-499.

83. Compton DA, Szilak I, Cleveland DW. Primary structure of NuMA, an intranuclear protein that defines a novel pathway for segregation of proteins at mitosis. J Cell Biol 1992; 116:1395-1408.

84. Yang CH, Lambie EJ, Snyder M. NuMA: An unusually long coiled-coil related protein in the mammalian nucleus. J Cell Biol 1992; 116:1303-1317.

85. Zeng C, He D, Brinkley BR. Localization of NuMA protein isoforms in the nuclear matrix of mammalian cells. Cell Motil Cytoskel 1994; 29:167-176.

86. Kallajoki M, Harborth J, Weber K et al. Microinjection of a monoclonal antibody against SPN antigen, now identified by peptide sequences as the NuMA protein, induces micronuclei in PtK2 cells. J Cell Sci 1993; 104:139-150.

87. Compton DA, Cleveland DW. NuMA is required for the proper completion of mitosis. J Cell Biol 1993; 120:947-957.

88. Compton DA, Cleveland DW. NuMA, a nuclear protein involved in mitosis and nuclear reformation. Curr Opin Cell Biol 1994; 6:343-346.

89. Zeng C, He, D, Berget SM et al. Nuclear-mitotic apparatus protein: a structural protein interface between the nucleoskeleton and RNA splicing. Proc Natl Acad Sci USA 1994; 91:1505-1509.

90. Blencowe BJ, Nickerson JA, Issner R et al. Association of nuclear matrix antigens with exon-containing splicing complexes. J Cell Biol 1994; 127:593-608.

91. Sibon OCM, Cremers FFM, Boonstra J et al. Localisation of EGF-receptor mRNA in the nucleus of A431 cells by light microscopy. Cell Biol Internat 1993; 17:1-11.

92. Hozak R, Sasseville AM-J, Raymond Y et al. Lamin proteins form an internal nucleoskeleton as well as a peripheral lamina in human cells. J Cell Sci 1995; 108:635-644.

93. Sparks CA, Bangs PL, McNeil GP et al. Assignment of the nuclear

mitotic apparatus protein NuMA gene to human chromosome 11q13. Genomics 1993; 17; 222-224.

94. Tang TK, Tang CJC, Chao YJ et al. Nuclear mitotic apparatus protein (NuMA): spindle association, nuclear targeting and differential subcellular localization of various NuMA isoforms. J Cell Sci 1994; 107:1389-1402.

95. Compton DA, Luo CG. Mutation of the predicted p34(cdc2) phosphorylation sites in NuMA impair the assembly of the mitotic spindle and block mitosis. J Cell Sci 1995; 108:621-633.

96. Chan P-K, Chan W-Y, Yung BYM et al. Amino acid sequence of a specific antigenic peptide of protein B23. J Biol Chem 1986; 261:14335-14341.

97. Fields AP, Kaufmann SH, Shaper JH. Analysis of the internal nuclear matrix: oligomers of a 39KD nuclear polypeptide stabilized by disulfide bonds. Exp Cell Res 1986; 164:139-153.

98. Krohne G, Stick R, Kleinschmidt JA et al. Immunological localization of a major karyoskeletal protein in nucleoli of oocytes and somatic cells of *Xenopus laevis*. J Cell Biol 1982; 94:749-754.

99. Schmidt-Zachmann MS, Hügle B, Scheer U et al. Identification and localization of a novel nuclear protein of high molecular weight by a monoclonal antibody. Exp Cell Res 1987; 153:327-346.

100. Yang L, Chow EC, Chewcheng SB et al. Fine-structural observation of a nucleolar-nuclear matrix-lamina-intermediate filament system in transformed cells. Anticancer Res 1994; 14:1829-1832.

101. Baran V, Vesela J, Rehak P et al. Localization of fibrillarin and nucleolin in nucleoli of mouse preimplantation embryos. Mol Reprod Dev 1995; 40:305-310.

102. Takeuchi K, Turley SJ, Tan EM et al. Analysis of the autoantibody response to fibrillarin in human disease and murine models of autoimmunity. J Immunol 1995; 154:961-971.

103. Lubben B, Rottmann N, Kubicka-Muranyi M et al. The specificity of disease-associated anti-fibrillarin antibodies compared with that of HGCL2-induced autoantibodies. Mol Biol Rep 1994; 20:63-73.

104. Cardido A, Medina FJ. Subnucleolar location of fibrillarin and variation in its levels during the cell cycle and during differentiation of plant cells. Chromosoma 1995; 103:625-634.

105. Creancier L, Prats H, Zanibellato C et al. Determination of the functional domains involved in the nuclear targeting of nucleolin. Mol Biol Cell 1993; 4:1239-1250.

106. Roussel P, Hernandezverdun D. Identification of Ag-NOR proteins, markers of proliferation related to ribosomal gene activity. Exp Cell Res 1994; 214:465-472.

107. Dickinson LA, Kohwishigematsu T. Nucleolin is a matrix attachment region DNA binding protein that specifically recognizes a region with high base-unpairing potential. Mol Cell Biol 1995; 15:456-465.

108. Zaidi SHE, Malter JS. Nucleolin and heterogeneous nuclear ribonucleo-protein C proteins specifically interact with the 3' untranslated region of amyloid protein-procurser messenger RNA. J Biol Chem 1995; 270:17292-17298.

109. Kondo K, Inouye M. Yeast NSR1 protein that has structural similarity to nucleolin is involved in pre-ribosomal RNA processing. J Biol Chem 1992; 267:16252-16258.

110. Bray D. Cell Movements. New York: Garland, 1992.

111. Stossel TP. On the crawling of animal cells. Science 1993; 260:1086-1094.

112. Zigmond SH. Recent quantitative studies of actin filament turnover during cell locomotion. Cell Motil Cytoskel 1993; 25:309-316.

113. Fey EG, Wan KM, Penman S. Epithelial cytoskeletal framework and nuclear matrix/intermediate filament scaffold: three-dimensional organization and protein composition. J Cell Biol 1984; 98: 1973-1984.

114. Agutter PS. Models for solid-state transport: messenger RNA movement from nucleus to cytoplasm. Cell Biol Internat 1994; 18:849-858.

115. Georgatos SD. Towards an understanding of nuclear morphogenesis. J Cell Biochem 1994; 55:69-76.

116. Berrios M, Fischer PA. A myosin heavy-chain-like polypeptide is associated with the nuclear envelope in higher eukaryotic cells. J Cell Biol 1986; 103:711-724.

117. Schindler M, Jiang L-W. Nuclear actin and myosin as control elements in nucleocytoplasmic transport. J Cell Biol 1986; 102: 859-862.

CHAPTER 4

EVENTS AT THE PORE-COMPLEX

INTRODUCTION

The background covered in the first three chapters puts us in a position to review the current trends of experimental research on nucleocytoplasmic transport. Most contemporary studies are predicated on the assumption that 'transport' can be equated with 'binding to the nuclear envelope and translocation through the pore-complex', a semantic convention that materially affects the design of experiments and the interpretation of results (chapter 1). A less arbitrary use of 'transport' leads to interpretations that are sometimes less straightforward but may be more internally consistent. For instance, data from experiments designed to elucidate translocation have occasionally elicited references to '(intra) nuclear transport' and 'nucleoskeleton' in 1990s publications, betokening some admission of confusion about what is actually being investigated. However, the fact that a consensus of contemporary nucleocytoplasmic transport researchers presume that their main if not sole focus of interest is the pore-complex requires us to try to review the literature in this same light. The difficulties attendant on their perspective will become apparent as we proceed.

There is some commonality between protein import and RNA export, which was first indicated by the discovery that an individual pore-complex can be engaged more or less simultaneously in both processes.[1] It is now apparent that some pore-complex polypeptides are relevant to both RNA and protein translocation,[2,3] though it may be difficult to decide in practice whether a functionally significant mutation in one of these components directly impairs interaction with the transport substrate or, alternatively, perturbs translocation because of total or partial distortion of the pore structure,[4] or alteration of the pore's relationship with other supramolecular assemblies. In any event, it would seem artificial to separate discussions of protein import and RNA export at this stage, and we shall accordingly review both in the present chapter. However, our emphasis will perforce be on protein movement, because most current research focuses on the import of nucleus-targeted (karyophilic) proteins.

We shall begin with a brief survey of the main current experimental approaches in the field and the concomitant interpretation difficulties, and then proceed to an evaluation of our growing knowledge: the functional organization of the pore-complex, the distinctive features of cellular macromolecules that identify them as transport substrates, the process of directing substrates to the pore-complex, and the mechanism of translocation itself. Throughout our discussion the issue of what 'transport' denotes will be a persistent underlying theme.

SOME EXPERIMENTAL APPROACHES

Nucleocytoplasmic transport studies in vertebrates depend nowadays on one of three approaches.[5] The first involves microinjection of the candidate transportant into the cytoplasmic or more rarely the nuclear compartment of the intact cell, usually an amphibian oocyte or a cultured mammalian cell. The second involves permeabilization of the cell, usually with digitonin, prior to addition of the transportant. The third requires isolated nuclei to be resealed with egg extract and the transportant to be added to the suspension. Movements of the candidate transport substrate to the nuclear periphery and into or out of the nucleus are then followed by fluorescence microscopy. Usually a fluorescent label is attached to the substrate or immunofluorescence methods are applied at various times after the start of the experiment (intact cells have to be permeabilized before addition of the labeled antibodies). In some cases an electron opaque label such as colloidal gold is employed and electron microscopy is used rather than fluorescence microscopy as the 'tracking' technique. This approach is relatively cumbersome but has the advantage of superior resolution of substrate interactions with, inter alia, the pore-complex.[6]

The springboard of much contemporary experimentation, greatly increasing the popularity of the second (permeabilized cell) approach, was the discovery that nuclei cannot take up known karyophilic proteins in digitonin-treated cells unless factors lost during the permeabilization process are resupplied. This finding, which will be discussed later, has focused attention on two questions, viz. how the pore-complexes interact with the implied 'soluble' factors and what relevance these factors have to the mechanism of translocation. Much of what is now believed about the binding of transportants to the nuclear envelope and the translocation process itself has derived from studies of such factors, but it is in respect of these beliefs that the denotation of 'transport' is most significant, as we shall see.

Our understanding of the functional organization of the pore-complex has largely come from a quite different experimental approach, involving yeasts, mainly *Saccharomyces cervisiae* but also *Schizosaccharomyces pombe*. The advanced state of *Saccharomyces* genome sequencing and the increasingly detailed protein data library afford opportunities for comparing vertebrate polypeptide sequences with yeast gene products. Yeast homologues of the 'soluble' factors mentioned in the previous paragraph have been identified and a permeabilized spheroplast preparation has been used. At present, and despite evidence for the evolutionary conservation of the pore-complex, not least in respect of the reiterated motifs in component polypeptides described in chapter 3, homologies between vertebrate and *Saccharomyces* pore-complex components remain largely uncertain. Comparisons of the structure's biochemical morphology between the disparate taxa are consequently speculative; nevertheless yeast genetic techniques combined with biochemical methods have thrown considerable light on the way in which the yeast pore-complex is organized functionally. Mutations in pore-complex component genes may be either lethal, temperature-sensitive or viable and these characteristics provide insights into whether a particular polypeptide, or a given domain therein, is essential for cell survival and growth. The effects of overexpression of single pore-complex genes can also be investigated.

One genetic technique that has proved very fruitful in pore-complex studies involves synthetic lethal mutants, in which a viable mutation in one gene (which serves as a screen) becomes lethal when a related gene is also mutated. Synthetic lethal mutants have led to the identification of previously unknown pore-complex components, complementing biochemical studies in which (for example) immunoprecipitation by a monoclonal against one known pore-complex polypeptide generates a subcomplex comprising several proteins which are therefore inferred to associate in vivo. They have also provided many insights into the relationship between the components and organization of the pore-complex and various translocation processes.

Results obtained from simple in vitro systems using subcellular fractions such as resealed nuclear envelope ghosts are becoming increasingly difficult to relate to this present mainstream of research.[7,8] The easy but unproductive solution to this difficulty is to ignore the results, a negative policy which we shall try to avoid in this book. Karyophilic protein uptake by isolated cultured cell nuclei appears to have the specificity expected from in situ studies,[9] but it is not yet clear whether the transportants are only binding to the nuclear periphery rather than entering the nucleoplasm.

PRESUMPTION AND INTERPRETATION

When experimental approaches become fashionable the difficulties of interpreting results are often only half-recognized. Conflicts and controversies arise all the time, and are hallmarks of active science, but they sometimes disguise deeper questions about the validity of the consensus perspectives within the constraints of which such conflicts and controversies have their being.

Let us consider first the yeast genetic studies that are elucidating the functional organization of the pore-complex. Suppose a mutation in polypeptide A, a known pore-complex component, becomes lethal when combined with a mutation in polypeptide B. Several explanations are possible. First, B might also be a pore-complex component and there might be functional redundancy between A and B so that the cell is viable unless both are crippled. Second, normal nucleocytoplasmic transport and therefore cell survival might depend on a structural association between A and B which is maintained if one but not if both proteins are mutated. Notice that this second explanation requires B to be a pore-complex component only if it is presumed that the pore-complex is the sole part of the cell pertinent to nucleocytoplasmic transport. If this presumption is not made then B might, for example, be part of the nucleoskeleton or cytoskeleton. Third, A and B might function in series in two component steps of transport. Each mutated form functions subnormally and when both are mutated the combined effect is lethal. The same argument holds in this case;

B need not be a pore-complex component unless it is presumed that no other part of the cell has a significant role in transport. Fourth, it may be that A has a key role in protein import and B one in RNA export, and functional linkage between these two processes results in lethality only when both are impaired. The general point is that even when the limited consensus view of 'transport' is adopted there is room for debate about the interpretation of results; but this very fact might obscure the inference that when the definition of 'transport' is broadened, the range of possible interpretations is broadened concomitantly.

Now let us consider the 'soluble' factors required to reconstitute karyophilic protein uptake in permeabilized cells. To begin with, a partial or complete loss of activity during permeabilization does not mean that the lost factors are exclusively in the solution phase in situ, as is tacitly assumed by many authors. The loss of membrane structure and the alteration of concentrations (notably protein concentrations) in the liquid in immediate contact with cytoskeletal elements are likely to alter protein distributions (chapter 2). However, there is no doubt that addition of these factors during nuclear import experiments has a dramatic effect on the results. In their absence, most micrographs of a fluorescently labeled karyophile show a general distribution of fluorescence through the cytoplasmic space and virtual exclusion from the nucleus. In the presence of some or all of them, the label is concentrated first around the nuclear periphery and then within the nucleus itself.

The standard consensus interpretation of such data, which we shall discuss in more detail later, is that there are two kinds of soluble factors, one required for binding the transportant to the nuclear envelope and the other for translocating it through the pore-complex. Notice that this type of interpretation presupposes the solubility of the digitonin-extractable factors. If transport exclusively involves the nuclear envelope (pore-complex) and the transportant must first bind there, then what other status can the factors have? If migration through the cytoplasm and nucleoplasm are accepted as mechanistically significant parts of the overall process of transport then the interpretation may be quite different, however. One class of factors might be required not so much for binding at the nuclear periphery as for conveying the transportant there through defined cytoplasmic routes, in which case they might be structural components of these routes (albeit labile ones) rather than solutes in the liquid phase. The other class could as well be involved in removing the transportants from the inner faces of the pore-complexes and distributing them within the nucleus as in the translocation process per se.

These issues—the ways in which the interpretation of results and indeed the design of experiments depends on our presumptions about what 'transport' denotes—will become clearer as we proceed with a review of the field. We have made the general point in advance partly to reemphasize our central concern in this book and partly because a grasp of it is likely to influence the reader's appreciation of the primary literature.

FUNCTIONAL ORGANIZATION
OF THE PORE-COMPLEX

One of the best established features of yeast pore-complex organization is the existence of a subcomplex of M_r approximately 2.4×10^5 comprising the polypeptides NSP-1, NUP-49, NUP-57 and NIC-96.[10-15] NSP-1 was one of the first yeast pore-complex components to be identified and NIC-96 is a major component by mass.[12-17] Both NSP-1 and NUP-49 mutants are impaired in protein transport so these two polypeptides are assumed to play a role in this process.[14] NUP-49 may also function in RNA export.[18] The more recently-identified NUP-159 is cytoplasmically directed and also plays a part in protein import.[19] Two other dominant components, NUP-1 and NUP-2, are homologous to one another and are genetically linked with NSP-1 which is homologous to vertebrate p62.[10,13,17,20] NUP-1 mutants have pleiotropic effects, implicating this polypeptide in both protein and RNA transport as well as in the maintenance of normal nuclear envelope structure and in nuclear segregation; it has been suggested that NUP-1 might participate in linking the pore-complex to the nucleoskeleton or the nucleolus.[3,21] However, not all NUP-1 mutants are lethal, and it is generally supposed that the homology between this protein and NUP-2 betokens redundancy of function.

NUP-49 interacts functionally with NUP-133 and NUP-158, mutations in either of which cause nuclear accumulation of poly(A) + RNA and also clustering of pore-complexes on the nuclear surface.[18,22,23] This correspondence between perturbation of normal pore-complex distribution and abnormality of mRNA transport is striking. It is also observed in mutations of NUP-145,[24,25] and nuclear envelope deformations and pore-complex clustering occur when NUP-116 or NUP-100 are mutated; these polypeptides interact genetically through their GLFG repeat motifs.[10,11,13,25] NUP-100, 116 and 145 share an RNA binding motif that has a preferential affinity for poly(G),[24] recalling the poly(G) stimulation of part of the presumed mRNA translocation machinery in mammalian liver nuclear envelopes.[26] This functional relationship between pore-complex distribution, RNA binding and mRNA transport suggests that unless the pore-complexes are properly ordered, presumably by virtue of association with the lamina or with extrinsic skeletal structures in nucleoplasm or cytoplasm, then mRNA transport is impaired. It is true that in some mutants pore-complex clustering occurs without much accompanying nuclear poly(A)+ RNA accumulation[18,22,27] and that in others either the impairment of RNA export can be reversed while clustering persists[18] or vice-versa.[23] However, although these exceptions suggest that the connection is not direct and is not a simple cause-effect relationship, the results overall imply that correct pore-complex distribution is linked with normal poly(A)+ RNA export.

These studies have related individual polypeptide components of the yeast pore-complex to particular transport and other functions (e.g., NSP-1 to protein import) and have suggested that such components operate in situ not as independent entities but as parts of functional subcomplexes, perhaps reflecting the structural organization described in chapter 3. There is also evidence for redundancy amongst components, as in the case of NUP-1 and NUP-2. Finally, the yeast genetic studies have indicated that some pore-complex polypeptides interact with structures outside the nuclear envelope and that these interactions are significant in the maintenance of normal nuclear organization as well as in transport and other functions. These general conclusions might well apply to vertebrate pore-complexes; the extent of evolutionary conservation makes this at least plausible. However, the uncertainties about homologies between individual yeast and vertebrate polypeptides make most of the specific findings from yeast genetic studies incapable of reliable extrapolation to vertebrate cells at present.

TRANSPORT SIGNALS: SUBSTRATE RECOGNITION

What distinctive features of macromolecules mark them as substrates for nucleocytoplasmic transport? Karyophilic proteins normally possess at least one nuclear location signal (NLS), a targeting sequence that is part of the mature polypeptide, not of a pro-protein. A variety of NLSs have been identified but most of them fall into two broad classes.[28] One class, of which the paradigm example (the SV_{40} large T antigen) was the first NLS to be unequivocally identified and sequenced,[29] comprises sequences of three to five basic residues flanked by helix breakers, usually proline or glycine. The other class comprises bipartite sequences in which a pair of basic residues is separated from a three to five residue basic amino acid block by a 10 residue spacer; the paradigm example is *Xenopus* nucleoplasmin.[30,31] At the secondary structure level the two classes might share a short-helix conformation with at least one face cationic at physiological pH,[32] but there is no unequivocal evidence that any such common structural property plays a part in the interactions involved in nuclear import.

Evidence that a candidate peptide sequence is truly a NLS depends on: (a) showing that mutations of it result in cytoplasmic instead of nuclear location of the microinjected mutant protein and (b) demonstrating nucleus-targeting in fusion proteins made by introducing the normal NLS (but not the mutant one or the reversed sequence) into a protein that is normally extranuclear.[33] Attachment of a bona fide but not a mutant or reversed NLS to an inert substrate such as colloidal gold also ensures its uptake into the nucleus.[1,6] Fusion evidence of some sort is crucial for the unequivocal identification of a NLS. However, it must be emphasized that possession of at least one NLS does not guarantee the nuclear location of a polypeptide in vivo. The sequence might be reversibly inactivated by some kind of modification such as phosphorylation, or it might not be exposed in the native protein.[33] Indeed, a conformation change might render a NLS alternatively functional or cryptic, as when hormones bind to cytoplasmic glucocorticoid receptors and occasion their entry to the nucleus.[34] Also, other signals within the polypeptide such as membrane spanning sequences might override the influence of the NLS.[33] Therefore a NLS is not a sufficient condition for nuclear entry, though it probably ensures it in most cases. Nor is it a necessary condition in all cases, because a protein with no NLS might be transported by virtue of binding tightly to one that has a NLS, thus 'hitchhiking' into the nucleus,[35] though examples of this kind are probably rare. However, the uptake of calmodulin by nuclei might also be independent of any NLS;[36] glycoconjugate uptake by nuclei appears to be energy-dependent but does not involve a NLS;[37] and although some proteins seem to be able to use their NLS for nuclear export as well as import,[38] some shuttling proteins such as the HnRNP core polypeptide A_1 utilize sequences that are quite different from a standard NLS.[39]

Reservations notwithstanding, the significance of NLSs in most cases of karyophilic protein import is beyond doubt. The export of RNAs from the nucleus to the cytoplasm also depends on intramolecular signals, though associated shuttling proteins probably play a part in this process.[39] For example, mRNA export requires a mature 3' end[40] that at least in some cases seems to involve poly(A)[41,42] and it also requires an intact 5' methyl cap.[43,44] Export of tRNAs seems to require integrity of a conserved common sequence in the D and T loops, especially around residue G_{57}.[45] Export of at least some snRNPs also depends on the 5' cap.[46] Again, however, possession of these signals does not guarantee export. Their effects can be overridden if they are modified or rendered cryptic by ribonucleoprotein organization, or if the RNA is bound to an intranuclear structure in such a way that its migration to the nuclear periphery is prevented. Experimental confirmation of the status of an export signal should be isomorphic with that of a protein NLS. If the signal is removed or crippled then transport is abolished, and if it is incorporated into some otherwise nontransportable RNA then export ensues.[42]

Both conceptually and methodologically there is an exact symmetry between nuclear protein import and RNA export in respect of the signals involved. Usually, absence of signal means that the molecule is not a transport substrate; signals can be modified or rendered cryptic, and the assumption that the signals operate by interacting (albeit indirectly) with the pore-complex might or might not be valid.

NUCLEAR LOCATION SIGNALS AND PROTEIN IMPORT

In 1988 it became apparent that nuclear import of karyophilic proteins is at least a two-stage process (Fig. 4.1). The NLS was found to participate directly in a rapid ATP-independent event resulting in binding of the karyophilic protein to the nuclear envelope. This was found to be succeeded by a slower step requiring

ATP hydrolysis during which the transportant was distributed through the internum of the nucleus.[47,48] The first step was initially presumed to involve binding of the transport substrate to a nuclear envelope receptor, which proved to be an elusive quarry; the hunt revealed a number of candidates showing surprisingly high K_d values, in the order of 20-100 nM, for NLS-containing substrates.[49-53] However, the quest was diverted by the discovery, mentioned earlier in this chapter, that factors extracted from cells during permeabilization were essential for nuclear protein import.[53,54] It was found that there were at least two classes of such factors, relating to the two operationally distinguished stages of transport;[54] one class was necessary for transportant accumulation at the nuclear periphery, the other for occupation of the intranuclear space. We shall discuss the second class in a later section and confine our attention here to the first. It transpired that these 'binding' factors are not mere supplements to the nuclear envelope binding process, but in fact constitute the elusive receptor itself. Various names have been given to them. We shall use the generic label 'location signal binding factors' (LSBs) to denote both these karyophilic protein receptors and their functional analogs in RNA export.

The digitonin extraction results and various in situ labeling studies have shown that LSBs for karyophilic proteins are located in the cytoplasm as well as the nuclear envelope; indeed, some data in the 1980s suggested predominant association with cytoskeletal fibrils.[55,56] A LSB can be identified and purified by virtue of its capacity to bind a normal but not a reversed NLS on a protein or an affinity column[57,58] and confirmed as an LSB by its capacity to support nuclear import of transportants, in the presence of other requisite factors, in either permeabilized cells[58] or resealed isolated nuclei.[59] The proliferation of names for LSBs was a consequence of their almost simultaneous discovery by a number of laboratories exploring a variety of cell types. In *Xenopus* oocytes the known LSB consists of two polypeptides with M_r values around 60,000 and 90,000, known respectively as importin 60 and importin 90.[58,60,61] These have homologues in bovine erythrocytes, from which a pair of proteins of M_r 54-56,000, known as the NLS receptor, has been isolated[57] together with a protein of M_r 97,500 which is required to reconstitute nuclear envelope binding of the substrate-NLS receptor complex.[62] Importin 60 has a high percentage homology to the NLS receptor,[60] though it is

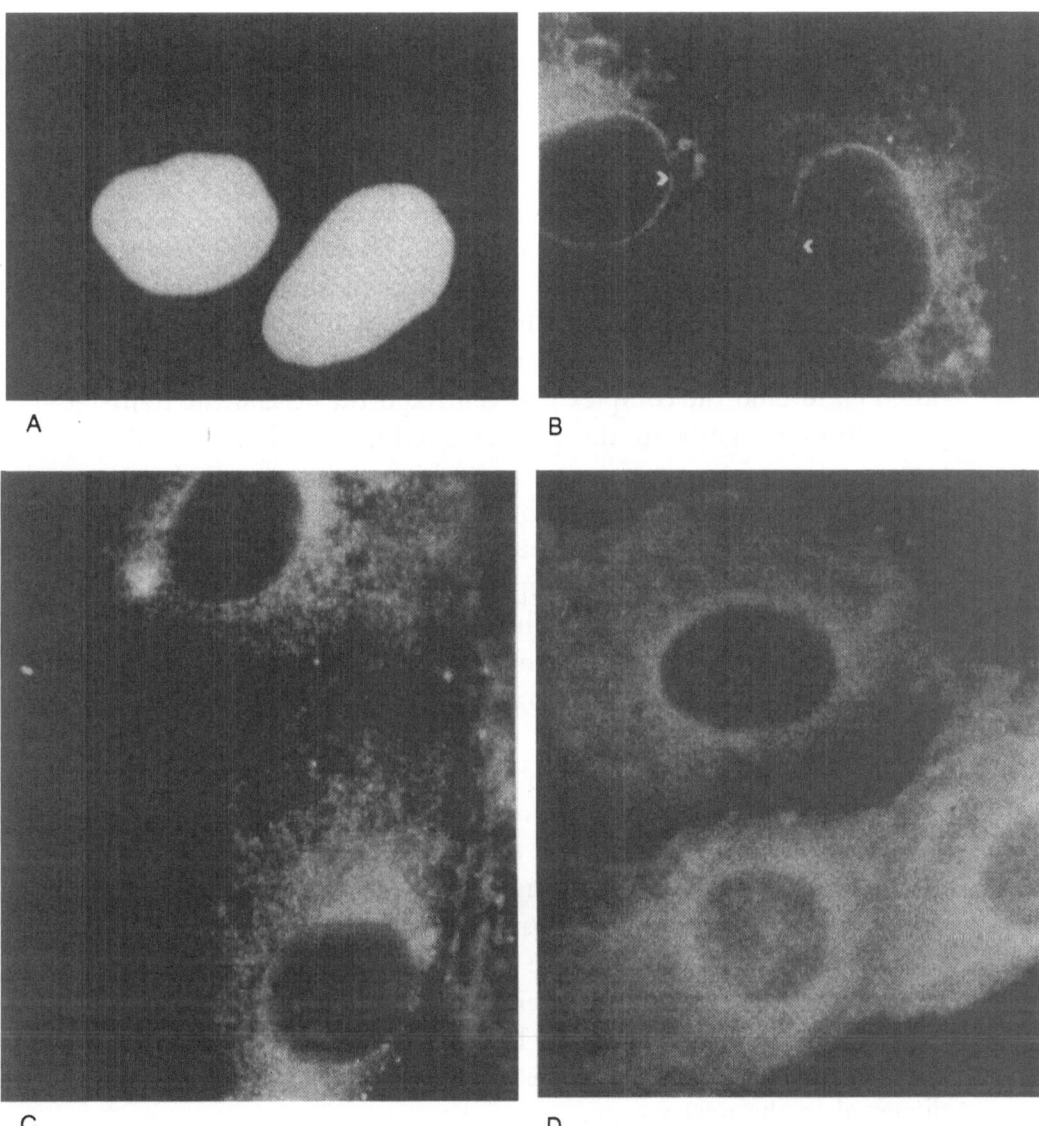

Fig. 4.1. Accumulation of nucleoplasmin at the nuclear envelope. Nucleoplasmin was injected into the cytoplasm of Vero cells, which were incubated as described below, then fixed and stained with mouse anti-nucleoplasmin antibodies followed by rhodamine-tagged rabbit anti-mouse and sheep anti-rabbit antibodies. Uninjected cells by comparison showed negligible fluorescence. (A) Incubation for 20 min at 37°C. Nucleoplasmin has accumulated within the nucleus. (B) Nucleoplasmin was microinjected as in (A), but at 20°C, and the coverslip was flooded with fixative immediately after microinjection of the last cell. The mean incubation time was less than 1 min at 20°C. A sharp perinuclear ring is evident (arrows). (C) The pentameric trypsin-resistant core of nucleoplasmin was microinjected and incubated as in (A). The nucleoplasmin core remains in the cytoplasm and appears evenly distributed without a sharp perinuclear fluorescence. (D) A control monoclonal IgG was microinjected and subsequently visualized with a rhodamine-conjugated anti-mouse IgG antibody. It shows a random cytoplasmic distribution. Reproduced with permission from Richardson WD et al, Cell 1988; 52:655-664. © Cell Press.

reputedly more sensitive to N-ethylmaleimide.[57-59] Importin 60 is also homologous throughout its length to the *S. cerevisiae* protein SRP-1[58,63] and to a variety of mammalian proteins including mouse and human SRP-1,[64] human RCH-1 protein[65] and the human influenza virus nucleoprotein NPI-1,[66] as well as to sequence fragments identified in a number of other species that include the pendulins from *Drosophila* and mouse.[60] The homology between human SRP-1 and bovine NLS receptor is particularly high.[62,64] Importin 90 is highly homologous with the M_r 97,500 factor from bovine erythrocytes and is believed to serve an equivalent function, i.e., to bind the complex between Importin 60 and the transport substrate tightly to the nuclear envelope.[60,67] It is also homologous to a variety of proteins, again functionally similar, in yeasts and human cells. The LSB identified in mouse ascites cells comprises a similar pair of polypeptides[68] and the karyopherins a_1, a_2 and b are once again homologous in structure and function.[69] The higher M_r component is tightly associated with both faces of the pore-complex and the binding is apparently dependent on at least one zinc ion, which is probably associated with the cysteine-rich domain.[70]

The general picture emerging from these studies is that LSBs comprise at least two types of polypeptide, one of M_r 50-60,000 that binds the NLS strongly and the nuclear envelope weakly, and one of M_r 80-100,000 that binds to the complex between its smaller counterpart and the substrate and markedly increases the affinity of this complex for the envelope. This image is shown schematically in Figure 4.2, where the two components are symbolized by LSB-A (smaller) and LSB-B (larger). Several candidate docking sites on the nuclear envelope have now been identified, as discussed in the next section.

INTERACTIONS BETWEEN PROTEIN TRANSPORTANTS AND LOCATION SIGNAL BINDING FACTORS

Human SRP-1 and RCH-1 (for example) are highly homologous but they are different gene products. This recalls the possibility raised during earlier investigations[52] that an individual cell might contain a family of LSBs differing in specificity for different NLSs and therefore different transportants. This would make the selection of proteins for nuclear uptake potentially amenable to elaborate choreography at the cytoplasmic binding stage of trans-

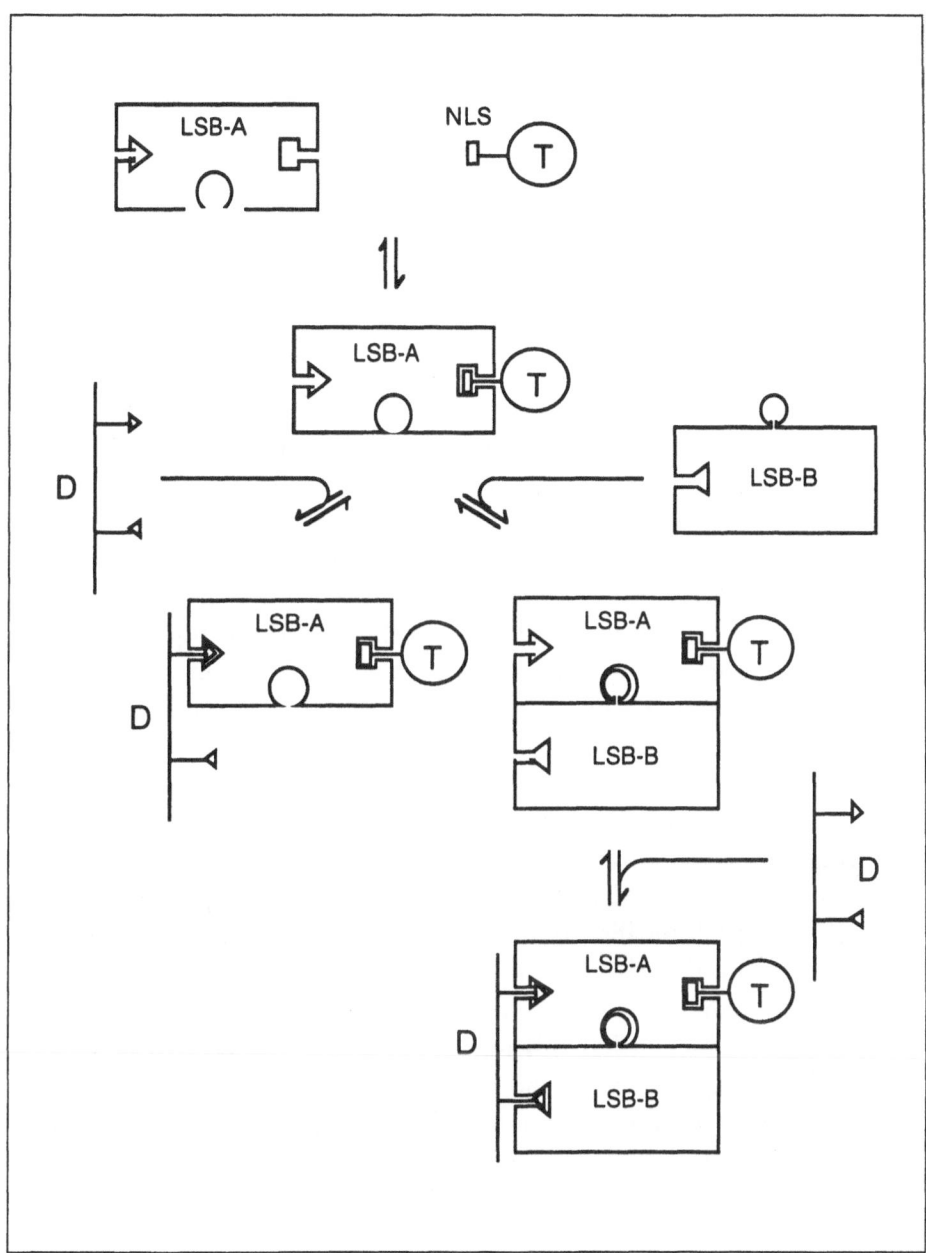

Fig. 4.2. Location signal binding factors and the attachment of karyophilic proteins to the nuclear envelope. The transportant binds to the LSB-A component (e.g., Importin 60), which can bind with relatively low affinity to the docking site. If LSB-B (e.g., Importin 90) is added then the affinity is greatly increased. This is speculatively interpreted here in terms of a two-component docking site, both components being used and the affinity correspondingly increased only if the ternary transportant-LSB-A-LSB-B complex is present. Key: T = Transportant; NLS = nuclear location signal; LSB-A = lower M_r component of location signal binding factor; D = docking site on nuclear envelope (pore-complex?); LSB-B = higher M_r component of location signal binding factor.

port but it requires NLS-LSB interactions to be subtler than simple electrostatic bonding,[49] which was dubious in any case in view of the inactivities of reversed-sequence NLSs.[19,52] Details of the NLS-LSB interaction are also interesting because they are pertinent to the physical state of the cytoplasmic LSB, notably to the assumption that LSBs are 'soluble' in situ.

The importins were first isolated from a complex of M_r $3\text{-}5 \times 10^6$ in high-speed supernatants from cell lysates,[58] which might or might not reflect their in situ organization. However, current evidence (see, for example, Fig. 4.2) suggests that the first identified step in nuclear import, envelope binding, entails the formation of at least a ternary complex. For an average sized transportant a complex comprising just one molecule each of transportant, LSB-A and LSB-B would have an M_r around 2×10^5 and the actual size might be considerably greater, as the relative mass of the isolated importin complex ($3\text{-}5 \times 10^6$) suggests.[58] For a complex of this size to migrate through the cytoplasm in a random, undirected way in the solution phase is somewhat implausible (chapter 2). For it to do so quickly—and we recall that the nuclear envelope binding step is rapid—is still less plausible. The belief that this migration takes place in the liquid phase therefore seems dubious. We might also recall some earlier data suggesting that LSBs associate with cytoplasmic fibrils;[55] and the fact that at least some LSBs have nuclear as well as cytoplasmic locations in situ[57,58] testifies to their capacity to bind to solid structures even in the absence of substrate. In goat uterus, the M_r 55,000 protein necessary for transferring estrogen receptors from cytoplasm to nucleus has inherent ATPase activity, though in view of earlier results it would be surprising if ATP were hydrolysed during transit of the cytoplasm.[71] These arguments are far from conclusive, but they do suggest that it is more natural to regard the first stage of karyophilic protein import not as binding from solution phase to nuclear envelope but as an ordered, mechanistically significant centripetal transcytoplasmic migration. The description of LSBs as 'soluble' factors might be no more than operationally valid.

The legitimacy or otherwise of this viewpoint has a bearing on our explanation for the stabilities of LSB-substrate complexes, but in any case some insights have been gained into the molecular details of the interactions involved. Both LSB components contain arm (armadillo) repeat motifs, 42 residue sequences that are predominantly hydrophobic and are believed to be involved in protein-protein binding.[61,72] The arm motif is widely distributed amongst cellular proteins, not all of which are obviously involved in nucleocytoplasmic transport.[72] Interestingly, the arms of SRP-1 are involved in binding to the recombination activating protein RAG-1 and the region of RAG-1 involved in this interaction contains a probable NLS.[64,65] It is not certain yet whether this NLS is functional or whether it participates directly in SRP-1 arm binding; but if these possibilities are realized then the range of cellular components potentially involved in nucleocytoplasmic exchanges is considerably extended. It becomes plausible that any polypeptide with an armadillo-like repeat can potentially bind an NLS-containing transportant, making it a LSB by definition.

LSB-substrate binding to the pore-complex is sensitive to wheat germ agglutinin[54,62] and therefore presumably involves one or more of the O-glycosylated pore components. This is probably not a mere steric effect, because in yeasts SRP-1 binds to the proteins NUP-1 and NUP-2 and less stably to NSP-1; the central xFxFG repeats seem to be involved in these associations.[3,15,20] Physical considerations suggest that LSB-transportant complexes will interact first with the cytoplasmic fibrils of the pore-complexes making NUP-180 and NUP-214 feasible guide-rails or perhaps even docking sites in vertebrates (chapter 3). However, the centrally located p62 is certainly involved in translocation[2,73,74] indicating that the LSB-transportant complex is transferred from the pore periphery to the p62-p58-p54-p45 assembly during or immediately prior to translocation. Yeasts might have a similar transfer of the complex from NUP-1/2 to NSP-1, but the molecular architecture remains unclear. Several nucleoporins are probably involved in docking. The higher M_r LSB component binds to the cytoplasmic domain of NUP-159 in yeast,[23,75] and NSP-1 is also involved in binding the transport complex.[76] In HeLa cells the very high M_r cytoplasmic ring nucleoporin, NUP-358, has been implicated in docking along with other transport-related functions.[77] Transfer of the docked complex to the P62-containing central channel probably requires a dimer of a small protein (subunit M_r 14,000) present in some 10^6 copies per cell,[71,78] and at least in the case of SV_{40} large T antigen a domain in the transportant itself (in this example the p53 binding domain) is required for opening the channel.[79]

It is likely that the LSBs themselves enter the nucleus during translocation. That they do enter and are subsequently reexported is suggested by several lines of evidence (see below), including the apparent role of SRP-1 in attaching yeast NUP-1 to the nucleoskeleton or the nucleolus,[3,21] and the uptake of at least the lower M_r component along with the transportant has been directly demonstrated by Laskey and his co-workers (the higher M_r component seems to be located on both pore-complex faces).[61] If this is the case then the mechanism by which the LSB-transportant complex is disassembled within the nucleus becomes pertinent. We shall return to this issue in the next section but one.

LSBS INVOLVED IN RNA EXPORT

RNA transport signals, like protein NLSs, bind to specific proteins as prerequisites for engagement with the rest of the transport machinery. Some of these factors shuttle between nucleus and cytoplasm as protein LSBs seem to do. For example, both protein import and mRNA export are associated with the shuttling of a M_r 36,000 HnRNP core polypeptide and of the 'heat shock' protein hsp70.[80-83] Both mRNA and snRNP export require proteins associated with the 5' methyl cap, and the cap binding proteins CBP50 and CBP20 have been specifically implicated.[43,44,84-87] A β-galactoside specific nuclear lectin, CBP35, seems to be co-transported with messengers to the cytoplasm[88] while a nuclear glucose-binding protein CBP67 might be involved in docking nuclear messengers on to a M_r 80,000 receptor in the envelope.[89] (Nomenclature here is unfortunate; the use of "CBP" to signify alternatively 'carbohydrate binding protein' and 'cap binding protein' does not contribute to clarity.) There is also evidence for the involvement of LSBs in tRNA and ribosome export.[87]

However, it is not clear whether these various factors are functionally important at the pore-complex as is often tacitly assumed or at an earlier stage of transport, i.e., during migration through the nucleus. Certainly the cap binding proteins are known to play a part in splicing,[87] which is intimately related to intranuclear transport (chapter 5). The protein P110, that binds oligo(A) or poly(A) and is implicated in mRNA export,[41,42] is found in the nuclear envelope where it appears to form part of the translocation machinery, but it also occurs in cytoplasmic fibrils[41] and may be equated with a protein identical in size and biochemical properties located in the nucleus.[90] The broad analogy between protein and RNA transport is sustained: in both cases receptors for the transport signals appear either to comigrate with the transportant or to form part of a widely-dispersed structural framework along which migration occurs, or both; interaction with the pore-complex constitutes only one part (at most) of LSB function, albeit a prominent one experimentally.

Some nucleoporins have been implicated in mRNA export. Apart from the involvement of P62 in translocation, which is well attested,[91] there must be docking proteins. In yeast these probably include NUP-145[92] and NUP-82;[93] deletions of the C-terminal domain of the latter causes nuclear accumulation of poly(A)+ RNA but no marked abnormality of the pore-complex morphology.[93] Vertebrate analogs have not yet been identified, but it will be interesting to know whether any nucleoporins can be shown to bind to cap-binding proteins[84] and whether the M_r 80,000 CBP67 binding component is a nucleoporin.[89] Components of this kind are probably confined to the pore-complex. However, in view of the evidence for substantial commonality in the factors involved in mRNA, tRNA, ribosome, snRNA and 5S RNA export from nuclei despite a measure of class specificity,[94,95] the analogy with protein import indicated in the previous paragraph seems to hold. The analogy may extend to the existence of 'negative' LSBs, cellular components that sequester macromolecules with transport signals and make them unavailable for migration across the nuclear envelope. The dominant cytoplasmic poly(A) binding protein seems to have this role in respect of poly(A)+ mRNA transport[42] and immobile proteins with armadillo repeats might function similarly in regard to protein transport (see above). This suggestion is, of course, purely speculative. Certain RNAs and RNPs also travel from cytoplasm to nucleus. This process appears to depend on NLSs active in RNA-associated proteins,[96,97] though sequences within the RNA itself might act as signals in some cases.[97,98]

RAN/TC4 AND ITS ANCILLARIES

The consensus interpretation is that LSB action is limited to transportant binding at the nuclear pore, but we have reasoned that LSBs also have a role in the distribution and migration of transport substrates within the cytoplasm and the nucleoplasm. We take a similar view of the other major class of 'soluble' factors required for normal nucleocytoplasmic transport. The consensus holds that just as LSBs are necessary for pore-complex binding, so the second class of factors is necessary for translocation.[53,54] We shall argue here that this second class also plays a part in intracompartmental (at least intranuclear) migration as well as, or perhaps instead of, translocation per se.

The second class of factors comprises the small GTP-binding protein Ran/TC4, its guanine nucleotide exchange factor RCC1, and a number of other functionally-interacting polypeptides (see below). The significance of Ran in nucleocytoplasmic transport was recognized when it was shown that karyophilic protein import in permeabilized cells required the reintroduction of this G-protein,[99,100] and was dependent on GTP hydrolysis.[101] Ran, like RCC1, was known to have a variety of other nuclear functions,[102-106] to which we shall return in chapter 5. Ran is the only known nuclear member of the Ras family, other members of which it resembles in M_r and guanine nucleotide binding domain sequence but not in the remainder of its primary structure.[103,104] RCC1 exists in several isoforms. It is a polypeptide of some 450 residues containing seven glycine-rich near-repreats.[105] Both Ran and RCC1 are highly conserved. In terms of molecular abundance, nuclear GTPases outnumber RCC1 by about 25-fold,[105] but RCC1 inactivation releases most of the GTPase activity to the cytoplasm.[106] Loss of RCC1 suppresses nuclear protein import,[107] and the inference from these studies is that RCC1 and Ran are involved in the maintenance of high-order nuclear structure and a range of nuclear functions of which protein uptake is only one.[102-107] Both factors are also required for RNA export[108-111] and protein export.[112,113] Interestingly, neither Ran nor RCC1 has been shown to contain a NLS.

The relative stability of the GDP and GTP bound forms of Ran, and the rate of GTP hydrolysis (which is intrinsically very low), depend on interactions with a variety of functionally related polypeptides.[114] The implications for control of nucleocytoplasmic transport are potentially interesting, because it is clear that GTP must be hydrolyzed for both normal protein import and poly(A)+ RNA export to occur.[115] One such ancillary polypeptide is the GTPase activating protein Ran GAP and its yeast homolog RNA1,[116,117] which synergizes with RCC1 in stimulating GTP hydrolysis.[117] Another is the inhibitory binding protein Ran BP1,[118] the smallest (M_r = 23,000) of a series of Ran binding proteins that have been identified, and the only one found in the cytoplasm.[119] RanBP1 and other binding proteins interact with Ran-GTP, not Ran-GDP, and inhibit RCC1-mediated nucleotide exchange. In the absence of Ran GAP they also inhibit GTPase activity, but in the presence of Ran GAP they further stimulate it.[119] Ran BP1 is another highly conserved protein[120] and essential to the normal functioning of Ran and RCC1.[121,122]

The interactions amongst Ran, RCC1, Ran GAP and the Ran BPs are kinetically and structurally complicated. RCC1 binds to both Ran-GTP and Ran-GDP with dissociation constants in the 10^5 M^{-1} range and in doing so it accelerates nucleotide dissociation by some six orders of magnitude.[123] The stimulation of GTPase activity by Ran BP1 in the presence of Ran GAP is dependent on the acidic C-terminal sequence-DEDDDL of Ran, which stabilizes GDP binding and allows Ran to interact with Ran BP1.[124] Ran-binding proteins contain a conserved, distinctive consensus sequence of some 150 residues,[125] part of which is homologous to sequences in nucleoporins including yeast NUP-2. The complexity of functional organization is increased by the discovery that RCC1 interacts not only with Ran and Ran BP1, but also with hsp70 (which is known to be a corequisite for mRNA transport; see earlier) and a large protein (M_r estimated at 340,000) in *Xenopus* extracts.[126] It may also interact with another identified 'soluble' component of the second factor class, placental protein pp15,[127] which is essential for transport in some cells.

Table 4.1. Effects of ancillary proteins on Ran activities

Ran	RCC1	Ran GAP	Ran BP1	GTPase Activity	Nucleotide Exchange
+	−	−	−	(+)	(+)
+	+	−	−	+	++
+	−	+	−	+	(+)
+	+	+	−	++	++
+	−	−	+	(+)	(+)
+	+	−	+	(+)	+
+	+	+	+	+++	++

present (+)/absent(−)

Most of the Ran binding proteins so far identified are nuclear and they resist extraction with Triton and high ionic strength solutions.[119] It is plausible that the largest of them, with M_r estimated at 200-300,000, is NuMA; we shall consider this idea further in chapter 5. However, it may be a nucleoporin. Several nucleoporins contain Ran-binding domains,[128] and there is good evidence that GTP is hydrolysed at the cytoplasmic face of the nuclear pore-complex by a mechanism involving the M_r 358,000 nucleoporin Ran BP2.[129,130] This protein might be the one identified by Bischoff et al[119] or the M_r 340,000 protein described by Saitoh and Dasso.[126] Ran BP2 or NUP-358 contains eight tandemly-repeated C_2C_2 type zinc fingers, multiple copies of a xFxFG repeat motif, three Ran-GTP binding domains and a leucine-rich region and it seems to be located in the cytoplasmic ring.[77,130] This structure and location might implicate Ran BP2 and GTP hydrolysis in the docking of LSB-protein complexes destined for nuclear import.[130] Nevertheless, GTP hydrolysis seems to accompany the appearance of karyophilic proteins in the nucleoplasm, not pore-complex binding.[53,54,99-101]

This body of data is equivocal. On the one hand the cytoplasmic face of the pore-complex contains nucleoporins capable of binding both LSB complexes and Ran-GTP: NUP-358 in vertebrates[77,128] and NUP-2 and NUP-159 in yeast.[19,75,76,122] This suggests that the two classes of 'soluble' factors could cooperate via one or a very few linking glycoproteins. On the other hand, it is intuitively unlikely that a process as complex and multifaceted as nucleocytoplasmic transport depends on so Spartan a molecular economy. The isoform of RCC1 that is undoubtedly necessary for transport seems to be chromosome-associated and most of its cooperating factors are intranuclear.[102-106] Normal oligo(A) dependent, wheat germ agglutinin sensitive, mRNA export from resealed nuclear ghosts occurs in the absence of Ran and ancillaries,[42] contrary to the hypothesis that Ran acts at the translocation stage. Inactivation of Ran or RCC1 restricts poly(A)+ RNA to the nuclear internum, not the nuclear periphery.[108,110,111] Intact LSB-protein complexes enter the nucleus and their disassembly and the subsequent recycling of LSB-A (and LSB-B?) may be the GTP requiring step.[61] Therefore, the weight of evidence seems to favor an intranuclear role for Ran and its accessories rather than involvement in translocation per se.[100] The data that originally led to the hypothesis of a translocation role[99-101] could readily be reinterpreted in this way if it were accepted that transport signifies more than pore-complex crossing, and this is the inference we provisionally draw. We suggest that Ran binding to the cytoplasmically-directed nucleoporins might indicate a step in the nuclear uptake of Ran. Alternatively it might be required for transfer of docked LSB complexes to the p62-containing central channel. If our interpretation is correct, this later alternative bestows a dual role on Ran: it facilitates the initial step in translocation of imported proteins, but it also (and principally) acts within the nucleus on karyophilic protein accumulation, LSB recycling, protein export and mRNA export.

THE TRANSLOCATION PROCESS

Figures 4.2 and 4.3(a) summarize the events that might immediately precede and succeed the translocation of karyophilic proteins, in the light of this argument. We conjecture that analogous events might attend the export of RNAs and RNPs; for instance, it seems that proteins that are candidate LSBs in RNA movement are recycled, implying cytoplasmic processes isomorphic with those proposed in Figure 4.3(a) but presumably involving factors other than Ran and its allies. But if the inference is accepted (contrary to the consensus of the first half of the 1990s) that Ran and RCC1 do not participate directly in translocation, we are left with the problem of what, if anything, can be asserted about the translocation process itself.

The first in situ evidence that nucleocytoplasmic transport was at least a two stage process depended on the demonstration that nuclear protein entry (though not envelope binding) was ATP dependent.[55,56] The apparent association of Ran with translocation has perhaps clouded this issue, implying that the ATP might serve merely to rephosphorylate GDP. However, it should be recalled that in vitro studies many years earlier had suggested that nucleocytoplasmic transport at least of mRNA comprised several stages[131] and even at the nuclear envelope stage, mRNA binding had been shown to be energy independent while translocation required ATP hydrolysis.[132] More recently, in situ evidence was also obtained for an ATP requirement in mRNA export[133] but the site of ATP utilization could not be inferred from these data, except that it was probably precytoplasmic. The simplest interpretation

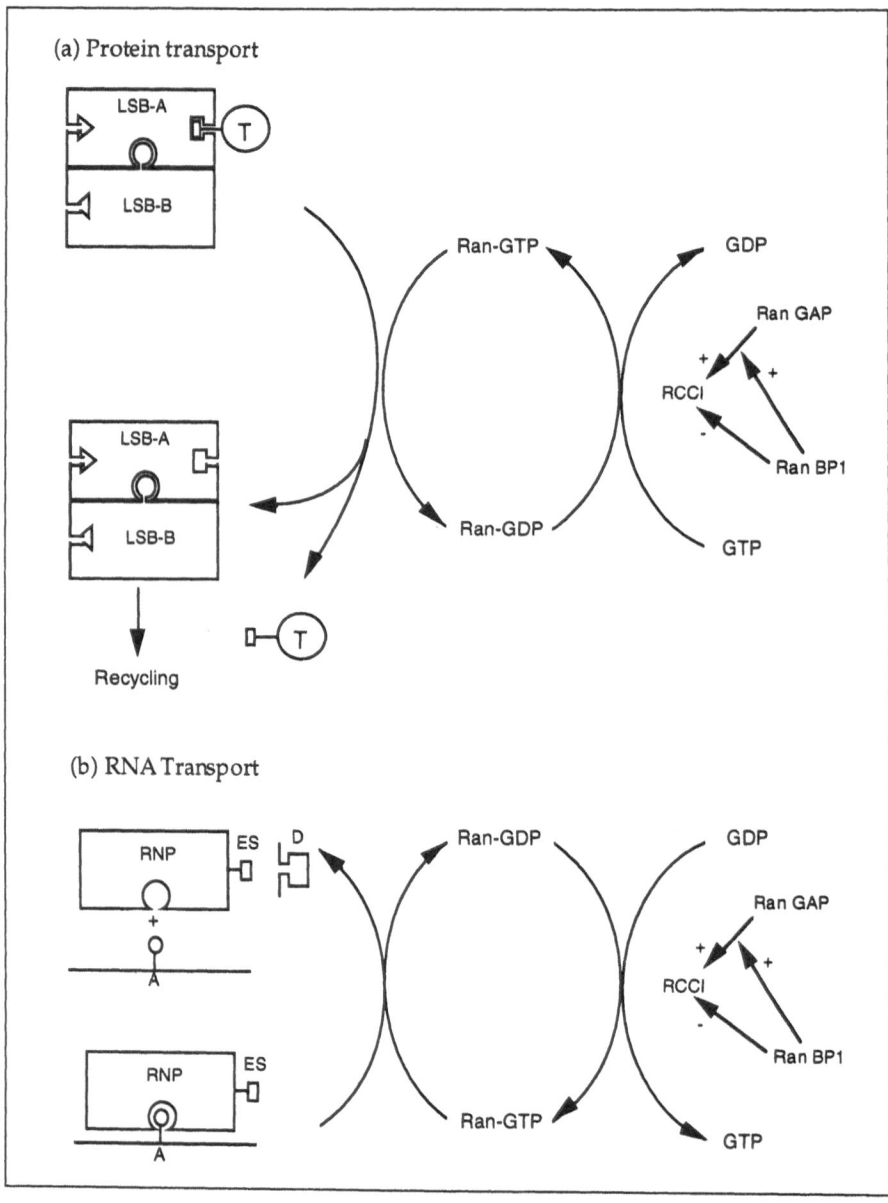

Fig. 4.3. Possible roles of Ran and RCC1 in nucleocytoplasmic transport. The hypothesis
proposed here is that GTP hydrolysis by Ran is used to separate (a) the karyophilic protein (T)
from the LSB, which is then available for recycling by export from the nucleus, and (b) the
ribonucleoprotein from its anchoring site(s) within the nucleus, making it available for binding
to the pore-complex and subsequent translocation. This proposal assumes that Ran and RCC1
act within the nucleus, not during translocation itself; the evidence seems to us more difficult
to reconcile with the consensus presumption. Key: (a) T = Transportant; LSB-A = lower M_r
component of location signal binding factor; LSB-B = higher M_r component of location signal
binding factor. (b) RNP = ribonucleoprotein; A = anchor (within nucleus); ES = export signal;
D = docking site (export signal binding site) on pore-complex.

makes use of the copious in vitro evidence for ATP hydrolysis by a nuclear envelope nucleoside triphosphatase as integral to mRNA export[41,42,131,132] but the possibility of further energy transduction at other stages of transport cannot be excluded, e.g., during migration within the nucleus.

Indeed, the nucleoside triphosphatase (M_r 40-46,000) may be associated not only with the inner nuclear membrane and presumably the pore-complex but also the nucleoskeleton.[131,134,135] Its connection with the rest of the translocating machinery has not been fully resolved, though the fact that the oligo(A) binding P110 and its proteolytic fragments are located on the inner face of the nuclear envelope,[136] as well as perhaps with the nucleoskeleton,[90] is suggestive as far as it goes. The relationship between P110 phosphorylation and mRNA affinity implies that the interaction between P110 and the nucleoside triphosphatase at the nuclear envelope level might be a regulatory step in transport.[131,136] One of the kinases involved appears to be protein kinase C, and phosphatidyl inositol hydrolysis products are likely to be generated in the inner nuclear membrane.[42,134,137] Kinetic models based on the assumption that mRNA transport is regulated exclusively at the nuclear envelope level give predictions in line with some but not all the relevant experimental data;[138] the occurrences of both the enzyme and the oligo(A) binding protein within the nucleus suggest that such simple models might not contain all the kinetically relevant factors.

The in vitro methods on which many of these findings are based have the advantage of being quantitative (in contrast, in situ immunofluorescence data are very difficult to quantify as a basis for kinetic models) and more readily interpretable in terms of, for example, sites of ATP utilization, because the systems used are compositionally much simpler than permeabilized cells. They seem to behave physiologically in terms of specificity, wheat germ agglutinin sensitivity and other criteria.[8,42,131] The difficulty of relating in vitro and in situ studies stems largely from a lack of critical comparability between them. For instance, investigators of mRNA transport in situ have failed to deploy antibodies against P110 and inhibitors of the nucleoside triphosphatase which would allow their results to be compared properly with in vitro findings. Conversely, investigators of the same process in vitro have not so far studied the role of the 5' cap and the cap binding proteins, which should certainly be done because in vitro methods hold out the promise of identifying the docking sites for cap binding proteins unequivocally. The present situation in which exponents of the two methodological traditions mostly neglect each others' contributions is unhelpful to say the least. It may be largely because of this that the genetic, architectural and biochemical relationships amongst p110, the nucleoside triphosphatase, cotransported proteins such as CBP35 and 67 and pore-complex polypeptides have not been adequately investigated or, still worse, even perceived as important questions by everyone in the field.

Whether the same nucleoside triphosphatase provides the energy for both mRNA and protein translocation is not clear. The fact that p62, and presumably the p62-p58-p54-p45 complex, has a role in both processes,[2,71,72] the role played by M_r 14,000 dimer in transferring important substrates to p62[78] and the lack of evidence for other ATP hydrolyzing enzymes in the nuclear envelope[131] all lend credence to the belief, but the question will not be answered until the functional relationships receive the ultrastructural and biochemical analysis that they merit. A possible approach to this problem lies in the observation that the pore-complex central channel changes from a 'closed' to an 'open' stage during translocation, a process that is intuitively likely to require energy and might be the ultrastructural manifestation of the nucleoside triphosphatase activity. In all translocation processes, the substrate appears to be transferred from the pore-complex periphery to the center of the lumen where the patency of an aqueous channel in the middle of the central plug changes markedly; this increase in channel diameter from 0.9 to 2.3 nm is the most striking difference between the 'closed' and 'open' states of the structure.[5,79,84,100,139-142] Presumably the initial binding of substrate to the periphery (see, for example, Fig. 4.2) accounts for the saturability of apparently all translocation processes and it may activate the ATP-hydrolyzing enzyme (as in the case of mRNA export) which inter alia opens the channel. This hypothesis is shown schematically in Figure 4.4; its heuristic potential is obvious, at least in general terms. It emphasizes that the docking sites for substrate or substrate-LSB complex need to be identified unequivocally and that the mechanochemical system involved in channel opening requires analysis.

It is here that the incomplete state of our knowledge of the biochemical morphology of the pore-complex becomes an obvious

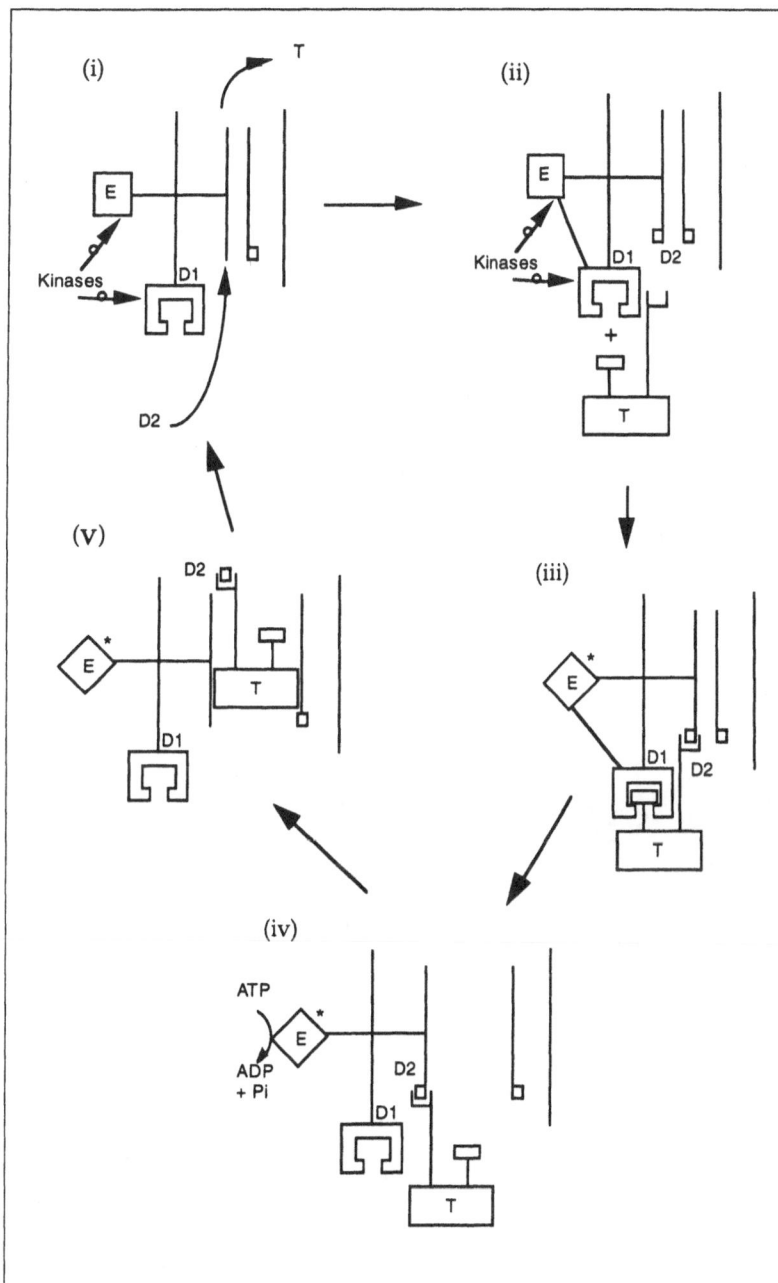

Fig. 4.4. A scheme for translocation. This view of translocation assumes that a transportant complex (T) has two docking sites in the pore-complex, one (D1) in the periphery and the other (D2) in the central channel; for instance, a poly(A)+ mRNA might bind via both an oligo(A) or poly(A) sequence and the 5' cap. There corresponds to the ternary complex shown in Figure 4.3 and D1 (and D2?) might accordingly have valencies greater than one. Both docking sites and the nucleoside triphosphatase (E) might be modulated by endogenous kinases. In (i) a transportant has just been translocated and the labile D2 site is being replaced. In (ii) a transportant is about to bind D1/D2 and in (iii) the bound transportant causes activation of the enzyme. The consequent hydrolysis of ATP (iv) opens the channel in the pore-complex center, and the transportant, detached from D1, is translocated through the resulting opening (v).

barrier. Until it is more nearly complete the proper synthesis of the various contributions of in situ and in vitro transport studies, genetic dissection, morphology, biochemical and immunological analysis and so forth will not be possible, and many aspects of the nuclear envelope stage of nucleocytoplasmic transport (still to some eyes the only significant stage) will remain obscure.

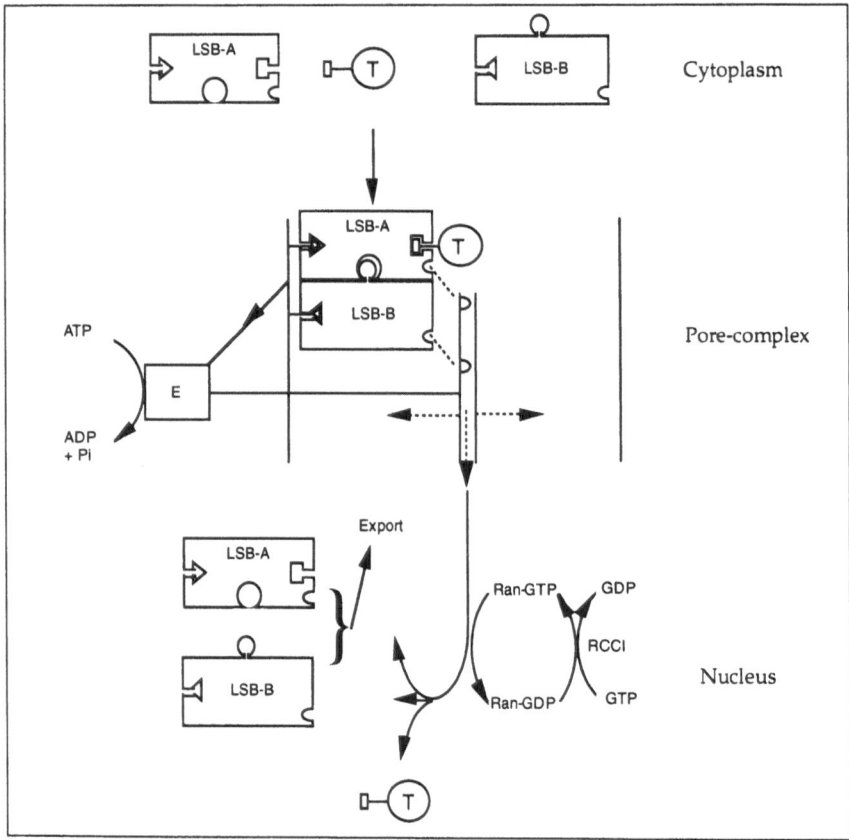

Fig. 4.5a. Protein import. Overview of events at the pore-complex. According to one possible interpretation of the current evidence, at least a ternary complex forms in the cytoplasm between the karyophilic protein (T) and two location signal binding proteins, one of which (LSB-A) binds the nuclear location signal, while the other (LSB-B) serves to stabilize interaction with a docking site (D1) on the periphery of the pore-complex. D1 is likely to be situated in the cytoplasmic fibrils; candidate components are NUP-180, 214 and 358 in vertebrates (possibly NUP-1/2 in yeasts). We speculate that occupancy of D1 by the transportant complex activates the ATP-hydrolyzing enzyme (E) that causes the LSB-transportant assembly to occupy docking site D2 and at the same time opens the central channel. D2 probably includes p62 in vertebrates and NSP-1 in yeasts. After translocation the ternary complex is disassembled by Ran/RCC1 and the liberated LSB components can be reexported.

OVERVIEW

This chapter is unlikely to have been easy reading even for these reasonably acquainted with the nucleocytoplasmic transport field. We have been obliged to juxtapose reviews of several disparate research lines; no one of these has yet generated a coherent picture, still less a simple one; and the underlying ambiguities of interpretation have added an extra layer of difficulty. We therefore append a composite scheme (Fig. 4.5) for the possible sequence of events at the pore-complex to summarize the tentative conclusions we have reached. Figure 4.5 should not be seen as a state-

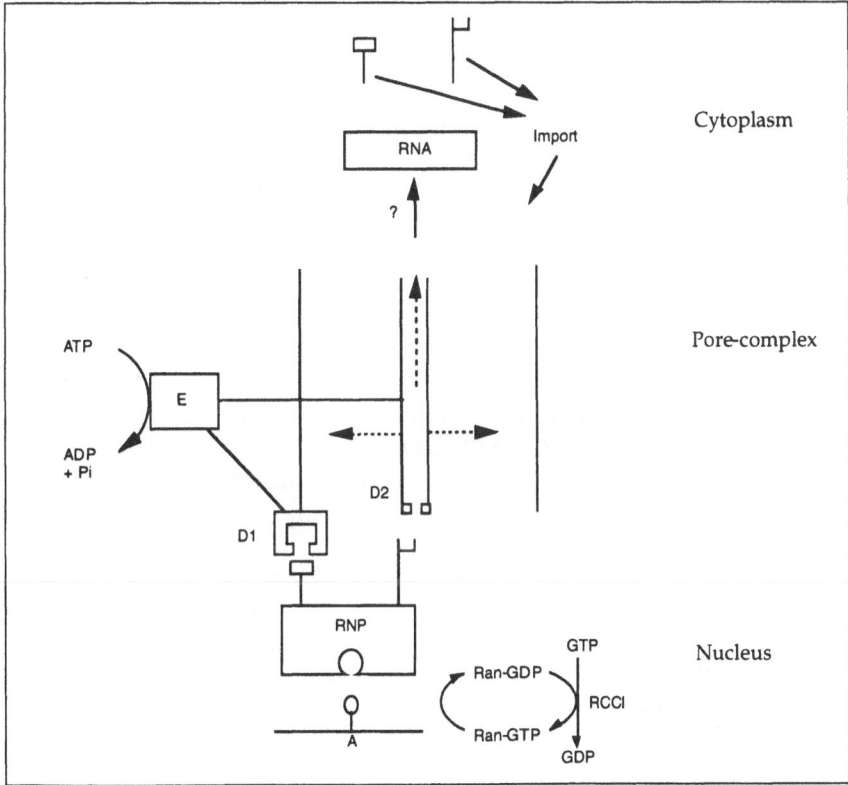

Fig. 4.5b. RNP export. The proteins severed from exportable ribonucleoproteins prior to translocation are assumed here to serve an anchoring role in the nucleus (A), and to be detached at the expense of GTP by the Ran/RCC1 system. This leaves the RNP available for binding to the peripheral (D1) and central (D2) docking sites on the pore-complex, occupancy of (the former of ?) which activates the nucleoside triphosphatase. The channel then opens and the RNP dissociates from D1 and is translocated. In the cytoplasm the components with presumed LSB function must somehow be detached so that they can be reimported, but to date there have been no insights as to how this occurs. Possibly the Ran associated with NUP-358 is involved.

ment of established fact, since each stage in it contains at least some element of speculation, but we hope that it will help to integrate the several themes of this chapter in the reader's mind. Although it should be regarded as no more than an aid to conceptualizing events at the pore-complex, it will serve to introduce our discussion of wider aspects of nucleocytoplasmic transport, which will be the topic of chapter 5.

REFERENCES

1. Dworetzky SI, Feldherr CM. Translocation of RNA-coated gold particles through the nuclear pores of hepatocytes. J Cell Biol 1988; 106:575-584.
2. Featherstone C, Darby MK, Gerace L. A monoclonal antibody against the nuclear pore-complex inhibits nucleocytoplasmic transport of protein and RNA in vivo. J Cell Biol 1988; 107:1289-1298.
3. Belanger KD, Kenna MA, Wei S et al. Genetic and physical interactions between SRP-1p and nuclear pore-complex proteins Nup1p and Nup2p. J Cell Biol 1994; 126:619-630.
4. Wente SR, Blobel G. A temperature-sensitive NUP116 null mutant forms a nuclear envelope seal over the yeast nuclear pore complex thereby blocking nucleocytoplasmic traffic. J Cell Biol 1993; 123:275-284.
5. Forbes DJ. Structure and function of the nuclear pore complex. Ann Rev Cell Biol 1992; 8:495-527.
6. Dworetzky SI, Lanford RE, Feldherr CM. The effects of variations in the number and sequence of targeting signals on nuclear uptake. J Cell Biol 1988; 107:1279-1288.
7. Riedel N, Fasold H. Transport of ribosomal proteins and RNAs. In: Feldherr CM, ed. Nuclear Trafficking. San Diego: Academic Press, 1992:231-290.
8. Agutter PS. Nucleocytoplasmic mRNA transport: a plea for methodological dualism. Trends Cell Biol 1994; 4:278-279.
9. Dean A, Kasamatsu H. Signal and energy dependent nuclear transport of SV40 Vp3 by isolated nuclei. J Biol Chem 1994; 269:4910-4916.
10. Wente SR, Rout MP, Blobel G. A new family of yeast nuclear pore complex proteins. J Cell Biol 1992; 119:705-723.
11. Wimmer C, Doye V, Grandi P et al. A new subclass of nucleoporins that functionally interact with nuclear pore protein NSP1. EMBO J 1992; 11:5051-5061.
12. Grandi P, Doye V, Hurt EC. Purification of NSP1 reveals complex formation with 'GLFG' nucleoporins and a novel nuclear pore protein NIC96. EMBO J 1993; 12:3061-3071.
13. Rout MP, Blobel G. Isolation of the yeast nuclear pore complex. J Cell Biol 1993; 121:771-783.

14. Schlenstedt G, Hurt EC, Doye V et al. Reconstitution of nuclear protein transport with semi-intact yeast cells. J Cell Biol 1993; 123:785-798.

15. Grandi P, Schaich N, Tekotte H et al. Functional interaction of Nic96p with a core nucleoporin complex consisting of Nsp1p, Nup49p and a novel protein Nup57p. EMBO J 1995; 14:76-87.

16. Hurt EC. A novel nucleoskeletal-like protein located at the nuclear periphery is required for the life-cycle of *Saccharomyces cerevisiae*. EMBO J 1988; 7:4323-4334.

17. Davis LI, Fink GR. The NUP1 gene encodes an essential component of the yeast nuclear pore complex. Cell 1990; 61:965-978.

18. Doye V, Wepf R, Hurt EC. A novel nuclear pore protein Nup133p with distinct roles in poly(A)+ RNA transport and nuclear pore distribution. EMBO J 1994; 13:6062-6075.

19 Kraemer DM, Strambio de Castillia C, Blobel G et al. The essential yeast nucleoporin NUP 159 is located on the cytoplasmic side of the nuclear pore-complex and serves in karyopherin-mediated binding of transport substrate. J Biol Chem 1995; 270:19017-19021.

20. Loeb JDJ, Davis LI, Fink GR. NUP2, a novel yeast nucleoporin, has functional overlap with other proteins of the nuclear pore complex. Mol Biol Cell 1993; 4:209-222.

21. Yano R, Oakes M, Tabb MM et al. Yeast SRP-1p has homology to armadillo/plakoglobin/β-catenin and participates in apparently multiple nuclear functions including the maintenance of the nucleolar structure. Proc Natl Acad Sci USA 1994; 91:6880-6884.

22. Li O, Heath CV, Amberg DC et al. Mutation or deletion of the *Saccharomyces cerevisiae* RAT3/NUP133 gene causes temperature-dependent nuclear accumulation of poly(A)+ RNA and constitutive clustering of nuclear pore complexes. Mol Biol Cell 1995; 6:401-417.

23. Gorsch LC, Dockendorff TC, Cole CN. A conditional allele of the novel repeat-containing yeast nucleoporin RAT7/NUP159 causes both rapid cessation of mRNA export and reversible clustering of nuclear pore complexes. J Cell Biol 1995; 129:939-955.

24. Fabre E, Boelens WC, Wimmer C et al. Nup 145p is required for nuclear export of mRNA and binds homopolymeric RNA in vitro via a conserved motif. Cell 1994; 78:275-289.

25. Wente SR, Blobel G. NUP145 encodes a novel yeast glycine-leucine-phenylalanine-glycine (GLFG) nucleoporin required for nuclear envelope structure. J Cell Biol 1994; 125:955-959.

26. Agutter PS, Harris JR, Stevenson I. RNA stimulation of mammalian liver nuclear envelope NTPase. Biochem J 1977; 162:671-679.

27. Pemberton LF, Rout MP, Blobel G. Disruption of the nucleoporin gene NUP133 results in clustering of nuclear pore complexes. Proc Natl Acad Sci USA 1995; 92:1187-1191.

28. Dingwall C, Laskey RA. Nuclear targeting sequences—A consensus? Trends Biochem Sci 1991; 16:478-481.

29. Kalderon D, Richardson WD, Markham AF et al. Sequence requirements for the nuclear location of SV_{40} large-T antigen. Nature 1984; 311:33-38.

30. Bürglin TR, de Robertis EM. The nuclear migration signal of *Xenopus laevis* nucleoplasmin. EMBO J 1987; 6:2617-2625.

31. Dingwall C, Robbins J, Dilworth SM et al. The nucleoplasmin nuclear location signal is larger and more complex than that of SV_{40} large-T antigen. J Cell Biol 1988; 107:841-850.

32. Agutter PS. Between nucleus and cytoplasm. London: Chapman and Hall, 1991.

33. Roberts, BC, Richardson WD, Smith AE. The effect of protein context on nuclear location signal function. Cell 1987; 50:465-475.

34. Picard D, Yamamoto KR. Two signals mediate hormone-dependent nuclear localization of the glucocorticoid receptor. EMBO J 1987; 6:333-3340.

35. Madsen P, Nielsen S, Celio JE. Monoclonal antibody specific for human nuclear proteins IEF 8Z30 and 8Z31 accumulates in nuclei a few hours after cytoplasmic microinjection of cells expressing these proteins. J Cell Biol 1983; 103:2083-2089.

36. Pruschy M, Ju Y, Spitz L et al. Facilitated nuclear transport of calmodulin in tissue culture cells. J Cell Biol 1994; 127:1527-1536.

37. Duverger E, Pellerin-Mendez C, Mayer R et al. Nuclear import of glycoconjugates is distinct from the classical NLS pathway. J Cell Sci 1995; 108:1325-1332.

38. Madan AP, de Franco DB. Bidirectional transport of glucocorticoid receptors across the nuclear envelope. Proc Natl Acad Sci USA 1993; 90:3588-3592.

39. Michael WM, Choi M, Dreyfuss G. A nuclear export signal in hnRNP A1: a signal-mediated, temperature-dependent nuclear protein export pathway. Cell 1995; 83:415-422.

40. Eckner R, Ellmeier W, Birnstiel ML. Mature mRNA 3' formation stimulates RNA export from the nucleus. EMBO J 1991; 10:3513-3522.

41. Schröder H-C, Diehl-Seifert B, Rottmann M et al. Functional dissection of the nuclear envelope mRNA translocation system: effects of phorbol ester and monoclonal antibody recognizing cytoskeletal structures. Arch Biochem Biophys 1988; 261:394-404.

42. Prochnow D, Thomson M, Benson M et al. Efflux of RNA from resealed nuclear envelope ghosts. Arch Biochem Biophys 1994; 312:579-587.

43. Hamm J, Mattaj IW. Monomethylated cap structures facilitate RNA export from the nucleus. Cell 1990; 63:109-118.

44. Hamm J, Darzinkiewicz E, Tahara SM et al. The trimethylguanosine cap structure of U1 snRNA is a component of a bipartite nuclear

targeting signal. Cell 1990; 62:569-577.

45. Tobian JA, Castano JG, Zasloff MA. tRNA nuclear transport: defining the critical regions of the human initiator tRNA-met by point mutagenesis. Cell 1985; 43:415-422.

46. Mattaj IW, De Robertis EM. Nuclear segregation of U2snRNA requires the binding of specific snRNP proteins. Cell 1985; 40:111-118.

47. Newmeyer DD, Forbes DJ. Nuclear import can be separated into distinct steps in vitro: nuclear pore binding and translocation. Cell 1988; 52:641-653.

48. Richardson WD, Mill AD, Dilworth SM et al. Nuclear protein migration involves two steps: rapid binding at the nuclear envelope followed by slower translocation through the nuclear pores. Cell 1988; 52:655-664.

49. Yoneda Y, Imamoto-Sonobe N, Matsuoka Y et al. Antibodies to asp-asp-glu-asp can inhibit transport of nuclear proteins into the nucleus. Science 1988; 242:275-278.

50. Adam SA, Lobl TJ, Mitchell MA et al. Identification of specific binding proteins for a nuclear location sequence. Nature 1989; 337:176-179.

51. Benditt JO, Meyer C, Fasold H et al. Interaction of a nuclear location signal with isolated nuclear envelope and identification of signal binding proteins by photoaffinity labelling. Proc Natl Acad Sci USA 1990; 86:9327-9331.

52. Yamasaki L, Lanford RE. Nuclear transport: a guide to import receptors. Trends Cell Biol 1992; 2:123-127.

53. Adam SA, Storne-Marr R, Gerace L. Nuclear import in permeabilized mammalian cells requires soluble cytoplasmic factors. J Cell Biol 1990; 111:807-816.

54. Moore MS, Blobel G. The two steps of nuclear import, targeting to the nuclear envelope and translocation through the nuclear pore, require different cytosolic factors. Cell 1992; 69:939-950.

55. Goldfarb DS, Gariepy J, Schoolnick G et al. Synthetic peptides as nuclear localization signals. Nature 1986; 326:641-644.

56. Breeuwer M, Goldfarb DS. Facilitated nuclear transport of histone H1 and other small nucleophilic proteins. Cell 1990; 60:999-1008.

57. Adam SA, Gerace L. Cytosolic proteins that specifically bind nuclear localization signals are receptors for nuclear import. Cell 1991; 66:837-847.

58. Görlich D, Prehn S, Laskey RA et al. Isolation of a protein that is essential for the first step of nuclear protein import. Cell 1994; 79:767-778.

59. Newmeyer DD, Forbes DJ. An N-ethylmaleimide-sensitive cytosolic factor necessary for nuclear protein import: requirement in signal-mediated binding to the nuclear pore. J Cell Biol 1990; 110:547-557.

60. Görlich D, Kostka S, Kraft R et al. Two different subunits of importin cooperate to recognize nuclear localization signals and bind them to the nuclear envelope. Curr Biol 1995; 5:383-392.
61. Görlich D, Vogel F, Mills AD et al. Distinct functions for the two importin subunits in nuclear protein import. Nature 1995; 377:246-248.
62. Adam EJH, Adam SA. Identification of cytosolic factors required for nuclear location sequence-mediated binding to the nuclear envelope. J Cell Biol 1994; 125:547-555.
63. Yano R, Oakes M, Yamagishi M et al. Cloning and characterization of SRP-1, a suppressor of temperature-sensitive RNA polymerase I mutations, in *Saccharomyces cerevisiae*. Mol Cell Biol 1992; 12:5640-5651.
64. Cortes P, Ye Z-S, Baltimore D. Rag-1 interacts with the repeated amino acid motif of the human homologue to the yeast protein SRP-1p. Proc Natl Acad Sci USA 1994; 91:7633-7637.
65. Cuomo CA, Kirch SA, Gyuris J et al. RCH-1, a protein that specifically interacts with the Rag-1 recombination activating protein. Proc Natl Acad Sci USA 1994; 91:6156-6160.
66. O'Neill RE, Palese P. NPI-1, the human homology of SRP-1, interacts with influenza virus nucleoprotein. Virology 1995; 206:116-125.
67. Chi NC, Adam EJH, Adam SA. Sequence and characterization of cytoplasmic nuclear protein import factor P97. J Cell Biol 1995; 130:265-274.
68. Imamoto N, Tachibana T, Matsubae M et al. A karyophilic protein forms a stable complex with cytoplasmic components prior to nuclear pore binding. J Biol Chem 1995; 270:8559-9565.
69. Morianu J, Blobel G, Radu A. Previously unidentified protein of uncertain function is karyopherin a and together with karyopherin b docks import substrate at nuclear pore-complexes. Proc Natl Acad Sci USA 1995; 92:2008-2011.
70. Chi NC, Adam EJH, Adam SA. Sequence and characterization of cytoplasmic nuclear import factor p97. J Cell Biol 1995; 130: 265-274.
71. Nirmala PB, Thampan RV. A 55 KDa protein (p55) of the goat uterus mediates nuclear transport of the oestrogen receptor. II. Details of the transport mechanism. Arch Biochem Biophys 1995; 319:562-569.
72. Peifer M, Borg S, Reynolds A. A repeating amino-acid motif shared by proteins with diverse cellular functions. Cell 1994; 76:789-791.
73. Davis LI, Blobel G. Identification and characterization of a nuclear pore-complex protein. Cell 1986; 45:699-709.
74. Baglia F, Maul GG. Nuclear ribonucleoprotein release and NTPase activity are inhibited by antibodies directed against one nuclear matrix glycoprotein. Proc Natl Acad Sci USA 1983; 80:2285-2298.

75. Radu A, Blobel G, Moore MS. Identification of a protein complex that is required for nuclear protein import and mediates docking of import substrate to distinct nucleoporins. Proc Natl Acad Sci USA 1995; 92:1769-1773.

76. Schlenstedt G, Hurt E, Doye V et al. Reconstitution of nuclear protein transport with semi-intact yeast cells. J Cell Biol 1993; 123:785-798.

77. Wu J, Mantunis MJ, Kraemer D et al. NUP 358, a cytoplasmically exposed nucleoporin with peptide repeats, Ran-GTP binding sites, zinc fingers, a cyclophilin A homologous domain, and a leucine-rich region. J Biol Chem 1995; 270:14209-14213.

78. Paschal BM, Gerace L. Identification of NTF2, a cytosolic factor for nuclear import that interacts with nuclear pore-complex protein p62. J Cell Biol 1995; 129:425-437.

79. Feldherr C, Cole C, Lanford RE et al. The effects of SV40 large T antigen and p53 on nuclear transport capacity in BALB/c 3T3 cells. Exp Cell Res 1994; 213:164-171.

80. Goldfarb DS. Shuttling proteins go both ways. Curr Biol 1991; 1:212-214.

81. Imamoto N, Matsuoka A, Kurihara T et al. Antibodies against 70-kD heat shock cognate protein inhibit mediated nuclear import of karyophilic proteins. J Cell Biol 1992; 119:1047-1061.

82. Piñol Roma S, Dreyfuss G. Shuttling of pre-mRNA binding proteins between nucleus and cytoplasm. Nature 1992; 355:730-732.

83. Shi Y, Thomas JO. The transport of proteins into the nucleus requires the 70-kilodalton heat shock protein or its cytosolic cognate. Mol Cell Biol 1992; 12:2186-2192.

84. Mehlin H, Daneholt B, Skoglund U. Translocation of a specific premessenger ribonucleoprotein particle through the nuclear pore studied with electron microscope tomography. Cell 1992; 69: 605-613.

85. Izaurralde E, Stepinski J, Darzynkiewicz E et al. A cap binding protein that may mediate nuclear export of RNA polymerase II-transcribed RNAs. J Cell Biol 1992; 118:1287-1295.

86. Izaurralde E, Lewis J, Gamberi C et al. A cap binding protein complex mediating Usn RNA export. Nature 1995; 376:709-712.

87. Izaurralde E, Mattaj IW. RNA export. Cell 1995; 81:153-159.

88. Agrwal N, Wang JL, Voss PG. Carbohydrate-binding protein 35. Levels of transcription and mRNA accumulation in quiescent and proliferating cells. J Biol Chem 1989; 264:17236-17241.

89. Schröder H-C, Facy P, Monsigny M et al. Purification of a glucose-binding protein from rat liver nuclei: evidence for a role in targeting of nuclear mRNP to nuclear pore-complexes. Eur J Biochem 1992; 205:1017-1026.

90. Schweiger A, Kostka G. Concentration of particular high mass phosphoprotein in rat liver nuclei and nuclear matrix decreases fol-

lowing inhibition of RNA synthesis with α-amanitin. Biochem Biophys Acta 1984; 782:262-268.

91. Dargemont C, Schmidt-Zachmann MS, Kühn LC. Direct interaction of nucleoporin p62 with mRNA during its export from the nucleus. J Cell Sci 1995; 108:257-263.

92. Fabre E, Boelens WC, Wimmer C et al. NUP145p is required for nuclear export of mRNA and binds homopolymeric RNA in vitro via a novel conserved motif. Cell 1994; 78:275-289.

93. Hurwitz ME, Blobel G. NUP82 is an essential yeast nucleoporin required for poly(A)+ RNA transport. J Cell Biol 1995; 130: 1275-1281.

94. Pokrywka NJ, Goldfarb DS. Nuclear export pathways of tRNA and 40S ribosomes include both common and specific intermediates. J Biol Chem 1995; 270:3619-3624.

95. Jarmolowsky A, Boelens WC, Izaurralde E et al. Nuclear export of different classes of RNA is mediated by specific factors. J Cell Biol 1994; 124:627-635.

96. O'Neill RE, Jaskunas R, Blobel G et al. Nuclear import of influenza virus RNA can be mediated by viral nucleoprotein and transport factors required for protein import. J Biol Chem 1995; 270:22701-22704.

97. Kambach C, Mattaj IW. Nuclear transport of the U2 snRNP-specific U2 B' protein is mediated by both direct and indirect signalling mechanisms. J Cell Sci 1994; 107:1807-1816.

98. Marshallsay C, Luhrmann R. In vitro nuclear import of snRNPs: cytosolic factors mediate m3G-cap dependence of U1 and U2 snRNP transport. EMBO J 1994; 107:1807-1816.

99. Moore MS, Blobel G. The GTP-binding protein Ran/TC4 is required for protein import into the nucleus. Nature 1993: 365:661-663.

100. Moore MS, Blobel G. A G protein involved in nucleocytoplasmic transport: The role of ran. Trends Biochem Sci 1994; 19:211-216.

101. Melchior F, Paschal B, Evans E et al. Inhibition of nuclear protein import by nonhydrolyzable analogues of GTP and identification of the small GTPase Ran/TC4 as an essential transport factor. J Cell Biol 1993; 123:1649-1659.

102. Kornbluth S, Dasso M, Newport J. Evidence for a dual role for TC4 protein in regulating nuclear structure and cell cycle progression. J Cell Biol 1994; 125:705-719.

103. Matsumoto T, Beach D. Premature initiation of mitosis in yeast lacking RCC1 or an interacting GTPase. Cell 1991; 66:347-360.

104. Dasso M. RCC1 in the cell cycle: The regulatory of chromosome condensation takes on new roles. Trends Biochem Sci 1993; 18:96-101.

105. Nishitani H, Ohtsubo M, Yamashita K et al. Loss of RCC1, a nuclear DNA-binding protein, uncouples the completion of DNA

replication from the activation of cdc2 protein kinase and mitosis. EMBO J 1991; 10:1555-1564.

106. Ren M, Drivas G, d'Eustachio P et al. Ran/TC4: A small nuclear GTP-binding protein that regulates DNA synthesis. J Cell Biol 1993; 120:313-323.

107. Tachibana T, Imamoto N, Seino H et al. Loss of RCC1 leads to suppression of nuclear protein import in living cells. J Biol Chem 1994; 269:24542-24545.

108. Kadowaki T, Goldfarb D, Spitz L et al. Regulation of RNA processing and transport by a nuclear guanine nucleotide release protein and members of the Ras superfamily. EMBO J 1993; 12:2929-2937.

109. Amberg DC, Fleischmann M, Stagljar I et al. Nuclear PRP20 protein is required for mRNA export. EMBO J 1993; 12:233-241.

110. Forrester W, Stutz F, Rosbash M et al. Defects in mRNA 3'-end formation, transcription initiation, and mRNA transport associated with the yeast mutation prp20: Possible coupling of mRNA processing and chromatin structure. Genes Dev 1992; 6:1914-1926.

111. Kadowaki T, Zhao, Y, Tartakoff A. A conditional yeast mutant deficient in mRNA transport from nucleus to cytoplasm. Proc Natl Acad Sci USA 1992; 89:2312-2316.

112. Guiochon-Mantel A, Delabre K, Lescop P et al. Nuclear localization signals also mediate the outward movement of proteins from the nucleus. Proc Natl Acad Sci USA 1994; 91:7179-7183.

113. Moroianu J, Blobel G. Protein export from the nucleus requires the GTPase Ran and GTP hydrolysis. Proc Natl Acad Sci USA 1995; 92:4318-4322.

114. Lounsbury KM, Beddow AL, Macara IG. A family of proteins that stabilize the Ran/TC4 GTPase in its GTP-bound conformation. J Biol Chem 1994; 269:11285-11290.

115. Schlenstedt G, Saavedra C, Loeb JDJ et al. The GTP-bound form of the yeast Ran/TC4 homologue blocks nuclear protein import and appearance of poly(A)+ RNA in the cytoplasm. Proc Natl Acad Sci USA 1995; 92:225-229.

116. Corbett AH, Koepp DM, Schlenstedt G et al. RNA1p, a Ran/TC4 GTPase activating protein, is required for nuclear import. J Cell Biol 1995; 130:1017-1026.

117. Klebe C, Bischoff FR, Ponstingl H et al. Interaction of the nuclear GTP-binding protein Ran with its regulatory proteins RCC1 and Ran GAP1. Biochemistry 1995; 34:639-647.

118. Hayashi N, Yokoyana N, Seki T et al. Ran BP1, a Ras-like nuclear G-protein binding to Ran/TC4, inhibits RCC1 via Ran/TC4. Mol Gen Genet 1995; 247:661-669.

119. Bischoff R, Krebber H, Smirnova E et al. Co-activation of Ran GTPase and inhibition of GTP dissociation by Ran-GTP binding protein Ran BP1. EMBO J 1995; 14:705-715.

120. Butler G, Wolfe KH. Yeast homologue of mammalian Ran binding protein 1. Biochim Biophys Acta 1994; 1219:711-712.

121. Ouspensky II, Müller UW, Matynia A et al. Ran-binding protein-1 is an essential component of the Ran/RCC1 molecular switch system in budding yeast. J Biol Chem 1995; 270:1975-1978.

122. Schlenstedt G, Wong DH, Koepp DM et al. Mutants in a yeast Ran binding protein are defective in nuclear transport. EMBO J 1995; 14:5367-5378.

123. Klebe C, Prinz H, Wittinghofer A. The kinetic mechanism of Ran-nucleotide exchange catalysed by RCC1. Biochemistry 1995; 34:12543-12552.

124. Richards SA, Lounsbury KM, Macara IG. The C-terminus of the nuclear Ran/TC4 GTPase stabilizes the GDP bound state and mediates interactions with RCC1, Ran GAP, and HTF9A/Ran BP1. J Biol Chem 1995; 270:10658-10663.

125. Ren M, Villamarin A, Shih A et al. Separate domains of the Ran GTPase interact with different factors to regulate nuclear protein import and RNA processing. Mol Cell Biol 1995; 15:2117-2124.

126. Saitoh H, Dasso M. The RCC1 protein interacts with Ran, Ran BP1, hsc70 and a 340 KDa protein in *Xenopus* extracts. J Biol Chem 1995; 270:10658-10663.

127. Moore MS, Blobel G. Purification of a ran-interacting protein that is required for protein import into the nucleus. Proc Natl Acad Sci 1994; 91:10212-10216.

128. Dingwall C, Kandelo-Lewis S, Seraphin B. A family of Ran binding proteins that includes nucleoporins. Proc Natl Acad Sci USA 1995; 92:7525-7529.

129. Melchior F, Guan T, Yokoyama N et al. GTP hydrolysis by Ran occurs at the nuclear pore-complex in an early step of protein import. J Cell Biol 1995; 131:571-581.

130. Yokoyama N, Hayashi H, Seki T et al. A giant nucleopore protein that binds Ran/TC4. Nature 1995; 376:184-188.

131. Agutter PS. Nucleocytoplasmic transport of mRNA: its relationship to RNA metabolism, subcellular structures and other nucleocytoplasmic exchanges. In: Müller WEG, ed. Progress in molecular and subcellular biochemistry, vol. 10. Heidelberg: Springer-Verlag, 1988:15-96.

132. McDonald JR, Agutter PS. The relationship between polyribonucleotide binding and the phosphorylation and dephosphorylation of nuclear envelope protein. FEBS Lett 1980; 116:145-148.

133. Dargemont C, Kühn LC. Export of mRNA from microinjected nuclei of *Xenopus laevis* oocytes. J Cell Biol 1992; 118:1-9.

134. Smith CD, Wells WW. Solubilization and reconstitution of a nuclear envelope associated ATPase; synergistic activation by RNA and polyphosphoinositides. J Biol Chem 1984; 259:11890-11894.

135. Clawson GA, Song Y-L, Schwartz AM et al. Interaction of human-immunodeficiency virus type 1 rev protein with nuclear scaffold nucleoside triphosphatase activity. Cell Growth Different 1991; 2:575-582.

136. Prochnow D, Riedel N, Agutter PS et al. Poly (A) binding proteins located at the inner surface of resealed nuclear envelope vesicles. J Biol Chem 1990; 265:6536-6540.

137. Schreifer P, Aitken SJM, Bachmann M et al. Immunological evidence for the localization of a 110 kDa poly (A) binding protein from rat liver in nuclear envelopes and its phosphorylation by protein kinase C. Cell Molec Biol 1993; 39:703-709.

138. Schröder H-C, Müller WEG, Agutter PS. Kinetic models for nucleocytoplasmic transport of messenger RNA. J Theoret Biol 1995; 174:169-177.

139. Feldherr CM, Kallenbach E, Schultz N. Movement of a karyophilic protein through the nuclear pores of oocytes. J Cell Biol 1984; 99:2216-2222.

140. Akey CW, Goldfarb DS. Protein import through the nuclear pore complex is a multistep process. J Cell Biol 1989 109:971-982.

141. Miller M, Park MK, Hanover JA. Nuclear pore complex: Structure, function, and regulation. Physiol Rev 1991; 71:909-949.

142. Akey CW, Radermacher M. Architecture of the Xenopus nuclear pore complex revealed by three-dimensional cryo-electron microscopy. J Cell Biol 1993; 122:1-19.

====== CHAPTER 5 ======

EVENTS INVOLVING
THE NUCLEOSKELETON
AND CYTOSKELETON

INTRODUCTION

In chapters 1 and 2 we argued that as a matter of principle processes within the nuclear and cytoplasmic compartments must be significant components of transport. In chapter 4 we found that assuming the opposite position, i.e., presuming that events at the nuclear envelope are the only ones relevant to nucleocytoplasmic transport, led to ambiguities and real interpretation difficulties. Evidently we must consider the roles of structures within the major cell compartments if we are to approach an adequate overall understanding, and here the discussion of 'characterization' in chapter 3 will become relevant. However, we must re-emphasize that evidence for (say) the involvement of a cytoskeletal element in a given transport process could be interpreted in many ways. Even if experimental artifacts are excluded, the fibrillar element might: (a) act as a 'railroad' for transport; (b) serve as an assembly site for transportant-containing complexes or as a processing or final anchoring site for them; (c) sequester transportants or corequisites by specific binding or nonspecific association; (d) exclude transportants or corequisites, confining them to aqueous channels; (e) generate fluid flow within such channels. These alternatives are not mutually exclusive and there may be others. Also, lack of free mobility in the cytoplasm does not necessarily imply (direct) association with cytoskeletal or membrane structures (chapter 2). In short, this aspect of nucleocytoplasmic transport is an interpretative minefield and it is advisable to proceed with caution. With regard to the nuclear compartment the situation is even more fraught because the nucleoskeleton is so much less well characterized than the cytoskeleton.

Our emphasis in chapter 4 was on proteins because more experimental data are available relating to karyophilic protein uptake through nuclear pores than to RNP (even mRNA) export. In this present chapter the emphasis will be different because much of the evidence relates to mRNAs and their precursors. Our conclusions will be uncertain and we shall be able to offer only tentative extrapolations to other categories of transportant. However, we shall be able to consider some of the questions raised in chapter 4 in greater depth, and begin to elucidate the most general issue of this book, the question of what 'nucleocytoplasmic transport' actually denotes.

MRNA AND THE CYTOSKELETON

In this and the following two sections we shall consider the role of the cytoskeleton in mRNA transport, and then in the following five sections address the even more difficult question of the role of the nucleoskeleton. It was shown in the 1960s that cytoplasmic mRNA does not re-equilibrate with nuclear material during open mitosis,[1] and later studies indicated that at least some mRNAs are rapidly immobilized when they are injected into the cytoplasm.[2] Removal of cytosolic components and membrane material from cultured cells by gentle extraction with detergent mixtures comprising nonionics and dilute deoxycholate left all translationally active polysomes bound to the cytoskeletal residue.[3,4] Viral messengers as well as host messengers were found to copurify with cytoskeletal preparations[5] and importantly, disruption of the microfilament system specifically inhibited translation. Cytochalasins B and D released messengers from the cytoskeleton in the concentrations needed to depolymerize F-actin.[6,7]

The simplest interpretation of these data is that all translationally active polysomes are attached to the actin cytoskeleton and that this attachment is a precondition of translation. This interpretation may be correct in many cases (see next section) but it is not unproblematic. For instance, messengers for proteins that are to be secreted from the cell or sorted into a membranous organelle are usually translated at the rough endoplasmic reticulum and their attachment to this membrane involves association between the nascent polypeptide and the integral signal recognition complex.[8] Perhaps such a messenger is also (independently) attached to an actin-containing two-dimensional protein gel contiguous to the endoplasmic reticulum (analogous to the spectrin-actin framework of the erythrocyte membrane or the lamina on the inner nuclear membrane).[3] This model would accommodate the data if the protein gel resists detergent extraction. However, it is also possible that the detergent extraction procedures did not remove all of the cytoplasmic membrane material, and in particular that they failed to remove the signal recognition complex-containing parts of the endoplasmic reticulum.[9] In vitro evidence is equivocal in any case because the enormous surface area of the cytoskeleton, in which the microfilament system is usually by far the most abundant component, affords opportunities for artifactual nonspecific binding of materials (not attached in vivo) when the cell is disrupted: the association of messengers with actin might appear specific simply because the cytoskeleton usually contains far more actin than anything else. Finally, subsequent experiments of the same kind have indicated that in most cells some 20-30% of the total cytoplasmic messenger population is readily extracted ('soluble') suggesting some heterogeneity of distribution.[9] We shall discuss the significance of this 'free messenger pool' in the next section but one.

On the other hand, there is supporting in situ evidence for the attachment of at least some active polysomes with the actin cytoskeleton and this attachment compartmentalizes certain translating messengers within different regions of the cytoplasm. Actin mRNA itself has been shown to be located specifically at sites of F-actin polymerization in some cells.[10-12] Microtubule associated protein mRNAs are anchored in neuronal dendrites[13] while tubulin, actin and glial acidic protein 3 messengers are restricted to the cell bodies.[14] These distributions seem to be functions of development; they are found in mature neurons but not in neurites.[15] In skeletal muscles, myosin mRNA is located near the sarcomeres.[16] When actin depolymerizing agents are added these distributions are randomized,[6,7] and this supports the view that messenger attachment to the actin cytoskeleton is a real and biologically significant phenomenon[17] and a prerequisite for translation.[3-5,18] However, these specific in situ findings apply mainly to messengers for cytoskeleton-related proteins, and the evidence for the greater generality of the phenomenon is largely in vitro.

In some systems, notably *Drosophila* and other insect oocytes, cytoplasmic subcompartmentalization of mRNA distributions appears to be a precondition for normal embryo development.[19] Some of these mRNAs are synthesized in the nurse cells[19,20] and appear to be imported into the oocyte not along microfilaments but along microtubules.[21,22] Microtubule motors seem to be involved in this transcellular migration, which has parallels in some vertebrate oligodendrocytes;[23] there is evidence for motor attachment involving a large domain in the 3' untranslated regions of the messengers,[24,25] though this domain is not necessarily involved in anchoring at the destination site.[26] Similar arguments apply to the apparently more widespread involvement of microfilaments in transcytoplasmic migration (next section): different mRNA-protein interactions might be involved in movement and final anchoring.

THE NATURE AND FUNCTION OF MESSENGER BINDING TO MICROFILAMENTS

In contrast to the established situation at the rough endoplasmic reticulum,[8] the polysomes are attached to F-actin or tubulin filaments in the cytoskeleton not via the nascent polypeptide chain but by the messenger itself.[6,7,24-26] It seems likely that at least some of this actin binding is attributable to the elongation factor eIF, which has been shown to be an actin binding protein in *Dyctostelium*[27] and may well be so in other taxa since its components are well conserved. Actin binding seems to reside in the eIF-4F complex, which contains both the cap-binding protein eIF-4E and the DEAD box RNA helicase eIF-4A. If actin association is required to activate this complex then the dependence of translation on cytoskeletal binding would in principle be explained, though the exact molecular mechanisms involved would remain unclear.

There are two pivotal points here:

1. If 'transport' is taken to end when a substrate (in this case a mRNA) reaches its site of utilization (in this case translation), then binding of the messenger via eIF and any other anchoring proteins involved constitutes the final step of transport, and we note that the location is

topologically remote from the pore-complex. In chapter 1, we reasoned that 'transport' must be understood in this way if we are to avoid semantic inconsistency and gratuitous fragmentation of our understanding of cell biology. It follows that the events involved in conveying the messenger from the pore-complex to the cytoskeletal binding site are also parts of nucleocytoplasmic transport.

2. If the model suggested by the *Dyctostelium* evidence is generally valid then both a DEAD box RNA helicase and a cap-binding protein are implicated in the ultimate cytoskeletal binding stage of mRNA transport. Since this stage involves fibril association, we might postulate that other RNA-fibril associations (at least RNA-F actin associations) also require RNA helicases and cap binding proteins. We shall return to this hypothesis later.

Although eIF-4E could and probably does serve to bring the 5' end of the messenger into contact with the helicase eIF-4A, it is likely that eIF-4B joins the complex before the RNA is unwound; eIF-4B has a probable RNA binding motif though one that might have no particular specificity.[29] Once the whole assembly is complete, eIF-4A hydrolyzes ATP and provides the energy for efficient unwinding of the messenger. It is possible that microfilament association is necessary to stabilize the multimolecular assembly. Alternatively, other factors could be involved that recognize specific sequences within the RNA and link it to a particular region of the cytoskeleton, thus providing a basis for the compartmentalization discussed in the previous section.

This model merits further examination; at present the evidence implicating the eIF complex in microfilament binding in vertebrate cells is only indirect. The idea is attractive but not established. The means by which a microfilament associated messenger progresses from the pore-complex to the ultimate cytoskeletal binding site also remains obscure, though there is immunological evidence for association of the oligo(A)-binding protein P110 with the cytoplasmic microfilaments as well as with the pore-complexes,[30] and of this same protein (P110) and its proteolytic fragments with cytoplasmic polysomes.[31] It is therefore possible that messengers bind via oligo(A) sequences such as the 3' poly(A) tail to the F-actin fibrils shortly after emergence from the pore-complex and that they are subsequently anchored at the 5' end via eIF. However, there is evidence that this is not the sole transcytoplasmic migration process (see previous section) and it may not even be the commonest one. In view of the capacity of P110-mRNA complexes in the pore-complex to stimulate the nucleoside triphosphatase (chapter 4) it might be thought that this enzyme is itself an RNA helicase, but the evidence here at least is clear; it is not.[32]

OTHER CYTOSKELETAL ATTACHMENTS AND THE SIGNIFICANCE OF THE 'FREE MESSENGER POOL'

The model advanced in the previous section is not only speculative but simplistic, for two main reasons: microfilaments are not the only anchors for cytoplasmic messengers; and for at least 20% of the total messenger mass the anchoring is at best insecure (see above). By considering these two issues, we can bring a more critical perspective to the model and assess some alternatives.

At least some mRNAs spend part of their cytoplasmic lifespans attached to microtubules[33] or intermediate filaments,[34] even in cell types where microtubules are not implicated in transport. However, even in the concentrations needed to disrupt intermediate filaments colchicine does not liberate active polysomes from such cells,[18,35] so it seems probable that messengers are not capable of being translated while bound to these portions of the cytoskeleton. In terms of the model we have advanced, this suggests that eIF activation requires actin, and other cytoskeletal proteins cannot substitute. Translationally inactive mRNA-containing particles have been identified, and although their in situ organization is not clear they might be associated with intermediate filaments in many cell types.[36] It is tempting to suppose that messengers newly emerged from the pore-complex associate with intermediate filaments before microfilaments or microtubules because the former elements usually have a dense perinuclear distribution and appear in many cases to be specifically anchored on the nuclear envelope.[37] These arguments suggest that there are at least two pools of bound or 'immobilized' mRNA in the cytoplasmic compartment in any given cell type, one (A) inactive and associated with intermediate filaments and (sometimes) microtubules, the other (B) active and associated with microfilaments or perhaps (in some cases) microtubules. Exchange and redistribution between these pools[38-40] might afford ways of controlling both the translation[9,36] and the degradation[40] of particular messengers. However, the existence of the 'free pool' (pool C)[9] must be recalled, and note taken of the fact that the size of pool C increases when the protein synthesis rate is increased.[38]

For simplicity, we neglect here the distinction between migrating and anchored messengers; pool B is assumed to contain both. We further assume that each pool A, B and C can exchange mRNA with the other two. Degradation is likely to be most rapid in C (because molecules less securely bound to fibrils are likely to be more accessible to RNAses) so the association between increase of C and the protein synthesis rate could be attributed to decreased degradation rate, which is a major factor in the kinetics of protein biosynthesis;[41] perhaps messengers in C are translatable, but perhaps a rapid increase in B by mass action accelerates translation. Similar arguments hold if C is directly entered by messengers leaving the nucleus; an acceleration of this process, e.g., in consequence of increased transcription,[41] will increase C and therefore the translation rate as before, either directly or via a mass-action increase in B or both (Fig. 5.1). If pools A or B rather than C are directly entered from the nucleus then the correlation between increase in C and the protein synthesis rate is less easy to understand, unless translational activity is largely confined to C, which would conflict with the evidence discussed previously in this chapter. It would be necessary to envisage a greater rate of flux from A to C than from C to B, making due allowance for changes in degradation rate, which would be surprising if B is a translationally active pool. Therefore, although topological considerations suggest A as the first postnuclear pool, kinetic considerations favor C for this role.

This leaves open the question of partitioning amongst the pools. The decision to translate or not to translate must entail the recognition of messengers for sorting to B (or C) on the one hand, A on the other. The signals needed to implement this decision have not been identified so far as we can see. Current evidence allows us to say the following, but no more:

  a. Excluding those associated with the rough endoplasmic reticulum, cytoplasmic messengers probably occupy at least three pools:[40] one, associated with microtubules and/or intermediate filaments, is untranslatable; of the others, one is associated with microfilaments and/or microtubules and one free. At least the former of these two comprises active polysomes.

b. Although kinetic and molecular details are obscure, messengers leaving the nucleus probably enter the 'free' pool before either of the others.

c. In the microfilament pool (pool B), messengers might be attached to the F-actin via (i) eIF, which binds the

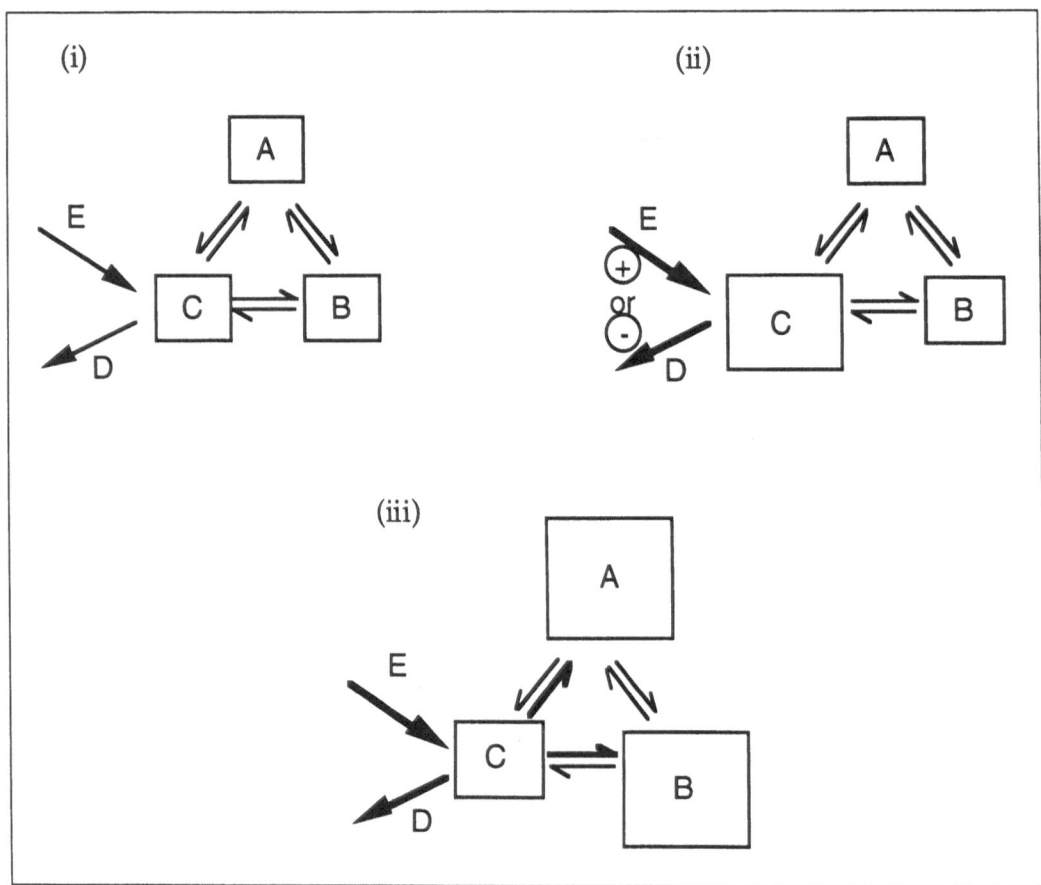

Fig. 5.1. Dynamics of cytoplasmic messenger pools. For simplicity we consider just three pools: A, cytoskeleton bound, translationally inactive; B, cytoskeleton bound, translationally active; C, not specifically bound, active or inactive. We propose that components of C are degraded (D) more rapidly than those of A or B and that messengers enter C directly on export (E) from the nucleus. This arrangement is illustrated in (i). In (ii) we note that either an increase in export (+E) or a decrease in the degradation rate (D) increases the size of pool C, and in (iii) that this will lead to increases in both A and B. Since the translation rate depends on the sizes of B (and C?) it follows that an increased protein biosynthesis rate correlates with an increase in the size of the 'free' pool C.

5' cap, (ii) P110 and its fragments, which bind oligo(A) fragments. In microtubule-dependent movement, association is probably through linkage between part of the 3' untranslated region and a motor. However, it has not been shown that these connections are general. Also, it is not clear whether pool B messengers in mature cells include only those encoding cytoskeletal and structurally associated proteins. Exchanges amongst the pools seem to correspond to changes in translation and mRNA degradation rates. Interpretation of these changes would be simplified if it were known whether pool C messengers included active polysomes.

On the other hand, it does seem clear that the cytoskeleton is involved in the partitioning and in the broadest sense of the term the transport of mRNA, and we can make some assertions about this involvement. The mere fact of compartmentalization implies that specific binding rather than nonspecific association occurs in at least some cases, and the generally small relative size of pool C implies that the cytoskeleton is not involved primarily as a molecular exclusion system or a fluid flow generator. What is not apparent from the available data is whether the cytoskeletal fibrils always provide transport tracks or only workbenches, e.g., for translation. This is analogous to the situation regarding karyophilic proteins and LSBs (chapter 4). The cytoskeleton is probably involved as an assembly site for the complexes but whether it also acts as a 'railroad' to the nuclear periphery, or merely as a way of defining 'canals' for such migration, is an open question.

MRNA, MRNA PRECURSORS AND THE NUCLEOSKELETON

We noted in chapter 3 that the nucleoskeleton almost certainly contains short F-actin fibrils; actin was established as a component of this elusive structure by several different laboratories.[42-44] If nuclear RNA is attached to the nucleoskeleton, then analogy with the cytoskeleton suggests that these actin fibrils might be involved in the attachment. Much of the early evidence for RNA association came from in vitro studies of isolated nuclear matrices, the physiological relevance of which was controversial, though the results did include the findings that: (a) spliceosome components copurified with the preparations; (b) exogenous RNA or RNP added during matrix isolation did not become firmly attached to the structures[45-52] and (c) there was at least some correspondence with in situ organization. The failure of oocyte nuclear RNAs to leak to the cytoplasm after nuclear envelope puncturing[53] and the association of transcription with the nuclear matrix/nucleoskeleton[47,54] lent credence to the belief. More recently, in situ studies have shown compartmentalization of specific transcripts within the nucleus, suggesting that, e.g., Epstein-Barr Virus transcripts[55] and c-fa transcripts in mouse fibroblasts[56] follow defined tracks from the transcription sites to the periphery. These arrangements survive chromatin depletion of the nuclei.[57] Interpretation of these data remains problematic[58,59] (see chapter 2), but if we also add that exogenous intron-containing premessengers associate with the discrete spliceosome-rich 'speckle' regions[60] then it seems fair to conclude that the weight of evidence favors some association between nuclear RNAs and organized actin-containing supramolecular structures that encompass splicing sites and are not part of the chromatin. Export of *Neurospora* RNAs from the nucleus is cytochalasin B sensitive in vivo.[61]

Evidence for the involvement of actin in the compartmentalization and movement of nuclear RNAs is both general and specific. Isolated rat liver nuclei containing $5\text{-}^{3}\text{H-U}$ prelabeled RNA can be suspended in media that maintain a close approximation to normal restriction in the presence of ATP,[62] but cytochalasin D promotes the large-scale nonspecific elution of acid-precipitable label under these conditions (Fig. 5.2). In nuclear matrices isolated from quail oviducts, cytochalasins specifically promote the liberation of incompletely-sliced ovalbumin messenger precursors.[63] Actin has also been shown by in situ studies to be associated with snRNPs and the splicing factor SC35, more intimately in differentiating cells.[64] However, the inference, although likely, cannot be considered certain. Cytoplasmic messengers have been shown to have more than one type of cytoskeletal attachment[40] (see above) and the situation is unlikely to be simpler in the nucleus. Association within in vitro preparations, however apparently specific and however close the resemblance of the preparation to an in situ structure, is always debatable evidence. Molecular details of the supposed attachment such as the identity of the actin binding proteins(s) involved are not available, and proximity to actin in situ could promote artifactual binding of an RNA that was

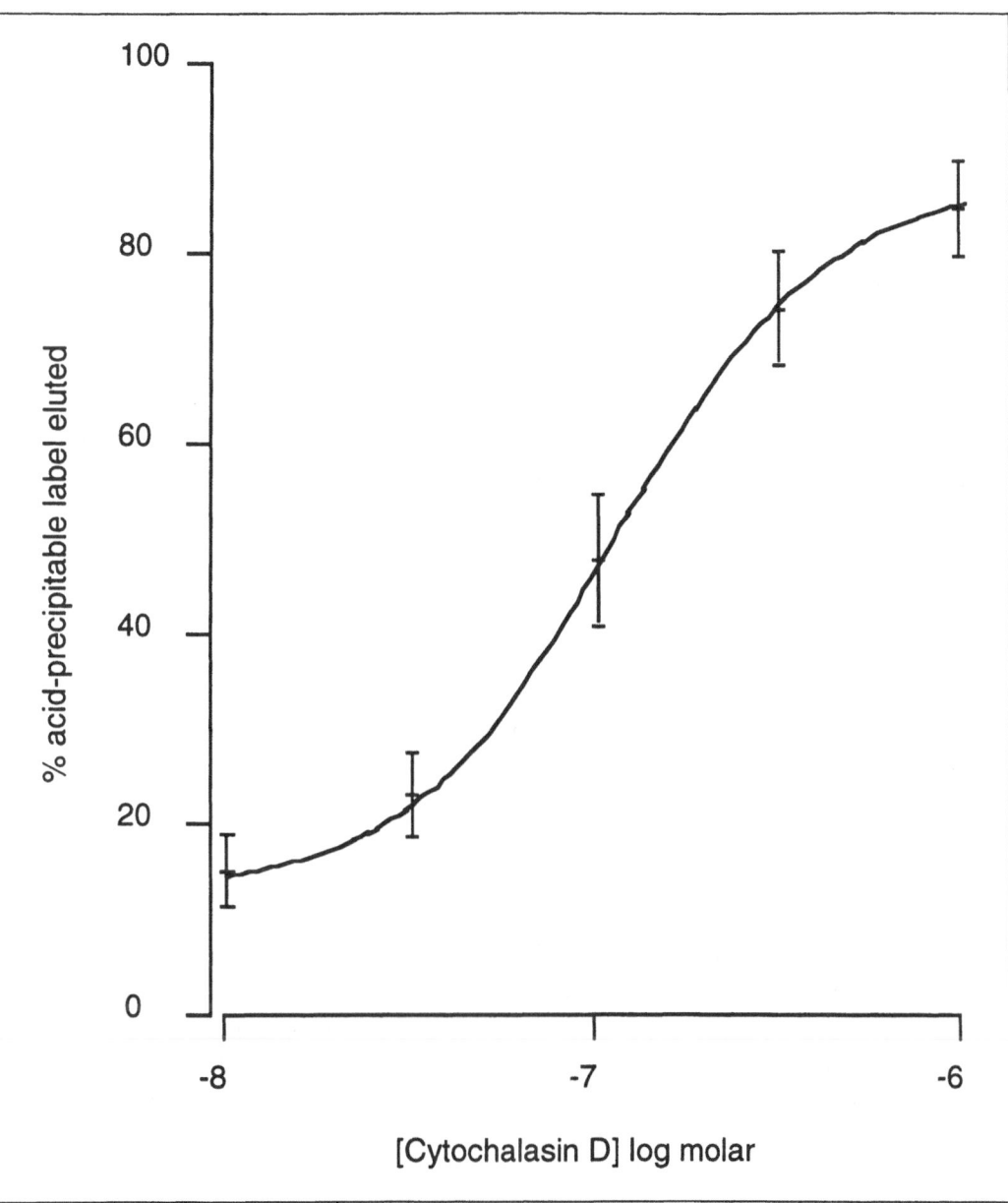

Fig. 5.2. Dose-response curve for RNA elution from isolated nuclei by cytochalasin D. Male Wistar rats (50-100 g body mass) were injected via the tail veins with 100 μCi 5-^{3}H-U. After 30 min they were killed by cervical dislocation and liver nuclei were isolated as described in ref. 90. Nuclei were suspended in concentrations of about 10^7 per ml in 25 mM Tris-HCl, 2.5 mM MgCl$_2$, 0.5 mM CaCl$_2$, 0.3 mM MnCl$_2$, 5 mM spermidine, 1 mM dithiothreitol, 1 mM vanadyl ribonucleotide complex, 1 mM PMSF, 2 mM ATP, pH 7.5 and incubated at 30°C for 5 min. Cytochalasin D was added in the concentrations specified. After centrifugation at 800 G for 5 min the eluted RNA was precipitated from the supernatants by the addition of 10% trichloroacetic acid. Radioactivity in the acid precipitates and in the nuclear pellets was measured as described.[61] Results are means ±sd of four experiments each performed in triplicate. Total nuclear radioactivity in these experiments was in the order of 10-12,000 dpm per 10^7 nuclei.

actually held in an 'extrachromosomal network' channel by molecular exclusion.[58] The fact that actin dissolution liberates RNAs could conceivably result from concomitant hydrodynamic changes in these channels. Finally, even if the greater weight of evidence is agreed to favor nucleoskeleton (specifically, actin) binding of nuclear RNAs, it cannot be inferred that all nuclear RNA is so immobilized; the existence of a 'free' pool in the cytoplasm makes a 'free' pool in the nucleus intuitively likely.[65] Here we must note the large dimensions of nuclear HnRNP particles and the narrowness of the extrachromatin network channels described by Zachar et al.[58]

In the case of the cytoplasm we were able more or less to exclude the possibility that the cytoskeleton acted negatively by confining most messengers to aqueous channels. In the case of the nucleus we cannot make the analogous statement with such certainty on the basis of the evidence discussed here. Even the tracks defined by specific transcripts[55-57] are not wholly persuasive because they are not shown by all premessengers[58] and an individual nascent transcript could, together with the encoding region of the chromosome, give a track-like appearance.[59] On the other hand, the fact that proteins can also shuttle along defined intranuclear tracks (the nucleolar protein Nopp 140 is the prime example) between nucleolus and pore-complex[66,67] lends further weight to the case for a positive role for the nucleoskeleton, with the pore-complexes serving as attachment sites for the track fibrils.[68,69]

SPLICING AND THE NUCLEOSKELETON

The association between actively transcribing genes and the nucleoskeleton has now become well established and three-dimensional analysis of chromosome organization suggests that active genes are probably located on the surfaces of chromosomal domains in the nucleus.[70,71] It follows that the nucleoskeleton interacts with the surfaces of chromatin regions and is therefore at least in very close proximity to nascent transcripts, since these are formed in immediate juxtaposition with it. This inference is clearly relevant to the apparently conflicting data (see above) about intranuclear RNA migration[57,58] though it does not resolve it.

During transcription, and apparently dependent on the process, snRNPs become associated with coiled bodies which contain the core M_r 80,000 protein coilin and the nucleolar protein fibrillarin as well as splicing factors.[72,73] These structures are therefore likely (in view of their composition and the transcription dependence of their assembly) to be nascent spliceosomes. Several points are pertinent here. First, the partially nucleolar origin of these assemblies is consistent with the observed role of the nucleolus in packaging poly(A)+ RNA and directing it towards the nuclear periphery,[74] as well as with the perinucleolar compartmentalization of at least some pre-mRNA processing events.[75] Second, there is some evidence for a close interfacing between NuMA and the spliceosome.[76,77] Third, since spliceosomes are not invariably clustered around the nucleolus and excluded from the remainder of the nucleus, there are obviously mechanisms either for moving fibrillarin and other coiled-body components around the nucleus preparatory to spliceosome assembly or for moving mature spliceosomes (the former seems intuitively more likely). Fourth, there is substantial evidence for the association of splicing with the nuclear matrix, consistent with earlier indications.[45,47,49,51,78] Since splicing is intimately coupled to transport, e.g., the removal of introns from genes that normally contain them can prevent transport to the cytoplasm of the transcripts,[79] and since it seems clear from the above discussion that transcription is structurally coupled with splicing as well as apparently with transport,[80] the involvement of the nucleoskeleton in this entire succession of processes seems to be well established. However, the nature of this involvement remains unclear.

Simplistic models have to be avoided. The cap binding complex involving CBP20 and CBP80 plays an important role in fitting the nascent transcript into the spliceosome and also in transport of messengers (whether through the nuclear internum or through the pore-complex remains unclear, though the latter has been tacitly assumed by some workers).[81,82] This might suggest that the very factors that facilitate splicing might have a role in transport, yet transport seems to depend on the removal of at least some splicing-related proteins from the RNP.[83] The role of actin in binding messenger precursors within the nucleus might, in view of the apparent role of RNA helicases in cytoskeletal binding, suggest the involvement of helicases at this level also (see above). However, the RNA helicases involved in splicing seem to interact with spliceosomes only transiently.[84,85] Further, the existence of actin fibrils in association with pre-mRNA suggests that myosins might be implicated in movement generation. Myosins have been demonstrated in nuclei [86] and suggested to have a role in transport,[43,86,87] but there is no specific evidence; for example, there is no clear evidence that spliceosomes contain myosin. On the other hand, it has been suggested that DNA topoisomerase II is a nucleoskeleton component and anchors mature (completely spliced) mRNA within the nucleus.[88,89] The role of DNA topoisomerase II during interphase has not become clear, and may be related to releasing strain induced in DNA helices during transcription and replication,[90] when the enzyme presumably associates with chromatin surfaces. This contrasts with its location in mitosis, when it seems to be integral to chromosomal cores, anchoring chromatin loops via AT-rich regions.[91] Since splicing also seems to be a prerequisite for export of RNAs other than messengers,[92] these mechanistic issues might be generalizable.

It might seem natural to return in this context to the role of Ran/RCC1, but for reasons that will become apparent we shall defer further discussion of these factors until later in the chapter.

PROTEIN TRANSPORT AND THE CELL SKELETONS

At this juncture it will be helpful to turn aside briefly from our mRNA-dominated discussion to ask again what roles the cytoskeleton and nucleoskeleton might play in the transport of karyophilic proteins. We reasoned in chapters 2 and 4 that proteins are no more likely to travel to and from the pore-complex by a process akin to 'diffusion' than they are likely to translocate through it by such a process. In any case, there is clear evidence that proteins are not generally freely mobile in either cytoplasm[93] or nucleus[94] and the nucleocytoplasmic concentration ratios of some karyophilic proteins cannot be accounted for by active transport across the nuclear envelope alone.[95] Earlier in the present chapter we mentioned examples of proteins that migrate along defined tracks within the nucleus,[67,68] and some published micrographs of fluorescently-labelled karyophilic proteins[96] suggest that they are concentrated along transcytoplasmic fibrils, the distribution of which would be consistent with the microfilament network. (Interpretation of such images is not straightforward, however.) In view of the repeated demonstrations that receptors for karyophilic proteins have wide intracellular distributions (LSBs; see chapter 4),[97-99] these lines of evidence suggest that karyophilic protein transport might involve some kind of solid-state mechanism.[100,101] Import of the estrogen receptor by goat uterus nuclei requires a 66,000 M_r LSB resembling importin (p66) and an acidic 55,000 M_r mediator (p55); of these proteins, p66 binds strongly to both tubulin and actin and p55 to actin.[102] Intranuclear binding of karyophilic proteins requires the NLS and also depends on prior passage through the pore-complex, perhaps suggesting association with an organized structure anchored and accessed via the pore-complex.[103] Intranuclear binding seems to be the main determinant of protein export processes,[104] and associations within compartments might contribute importantly to the behavior of shuttling RNAs and RNPs as well as proteins. (Nevertheless, nuclear targeting of cytoplasmic RNPs seems mechanistically akin to that of proteins,[105] and the comparability might also apply to nuclear protein export.) Overall, there is considerable support for the hypothesis that long-range intracompartmental structures play a part in protein transport.

The predominantly low K_d values for transportant-LSB complexes are consistent with this view. By analogy, cells interact with exogenous modulators such as hormones through small numbers of receptors with low K_d values but with extracellular matrix fibers through large numbers of receptors with high K_d values. The former type of interaction involves the solution phase and the latter is a form of solid-state transport, where (in the case of motile cells) receptors are occupied in succession with fairly rapid dissociation events at each step. Nucleocytoplasmic transport generally seems to involve large numbers of receptors with high K_ds.

One difficulty with this account of nuclear protein uptake is that cytoplasmic transport to the nuclear envelope is not only rapid but ATP-independent. Formal explanations for this can be offered: for example, that there is a succession of 'immobile' (perhaps microtubule or microfilament associated?) receptors for transportant or transportant-LSB complex where the binding is negatively cooperative and the relaxation to the high-affinity state is slow (Fig. 5.3). However, there is no evidence for or against this or any alternative model, not least because the relevant experimental data have not been sought. It is at least as plausible that the complexes are conveyed centripetally by fluid fluxes along the channels defined by cytoskeletal fibrils, though again the lack of ATP requirement has to be borne in mind: how would such fluid fluxes be generated?

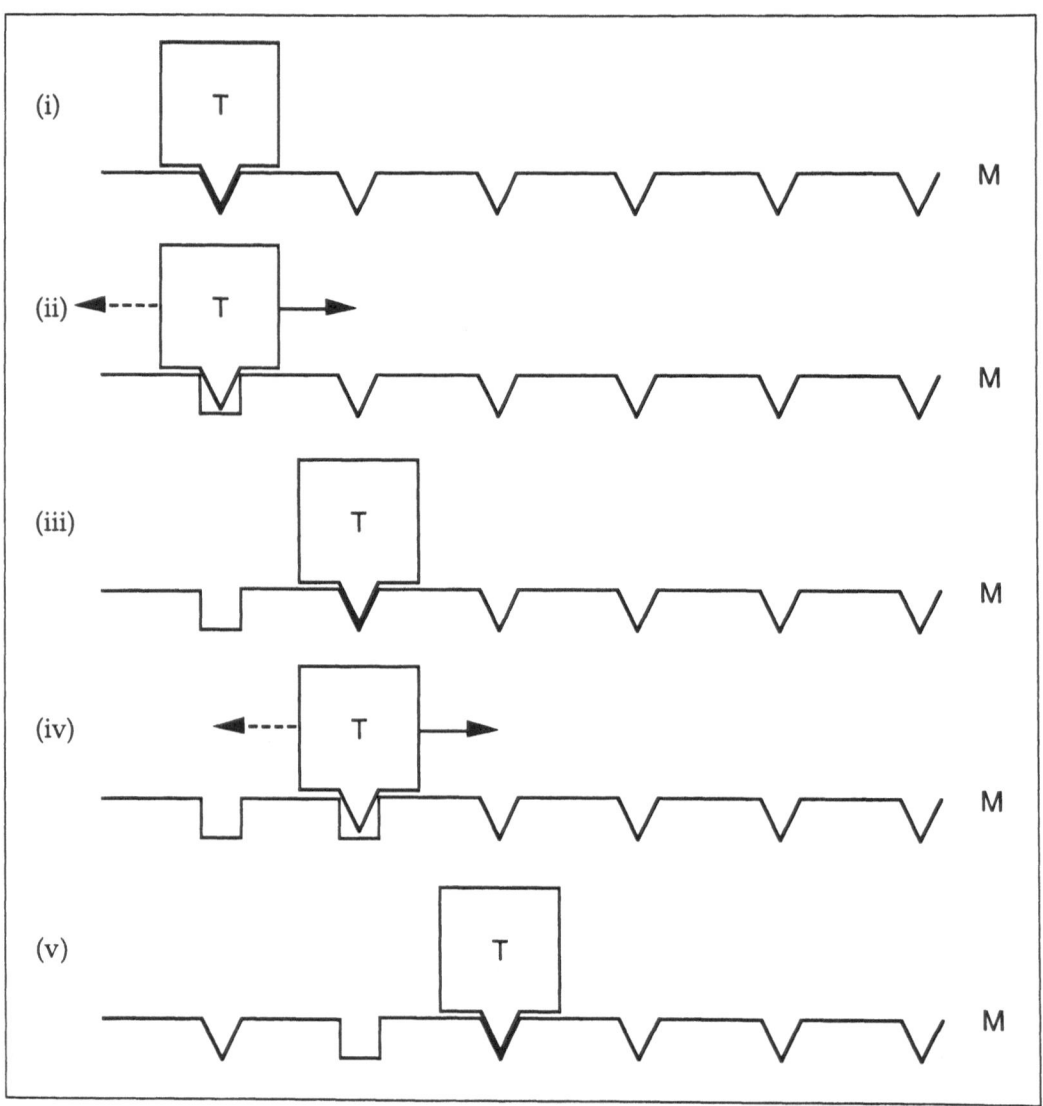

Fig. 5.3. ATP-independent directed transport in the cytoplasm? A transportant (T) could move along a 'solid state' migratory pathway (M) if the receptors on (M) were negatively cooperative. T binds to the receptor in its higher affinity state (i) and this changes the receptor to a low affinity state (ii), liberating T and allowing it to bind to the second receptor (iii). Here the same thing happens (iv), but T binds to the third receptor (higher affinity) rather than returning to the first, which does not relax from its lower affinity state until after T has moved on (v). This type of scheme produces uncomfortable feelings related to the second law of thermodynamics, though the energy for changing receptor state, like that for the binding and dissociation events, could perhaps be obtained from thermal sources, in which case the mechanism proposed here is a molecular analog of a vibrating conveyor.

Another difficulty is the interpretation of events within the nucleus. Suppose, as seems highly likely, the protein enters an aqueous space immediately after translocation and that this space is continuous with the extrachromosomal network channels.[58] Since these channels are narrow and, as we have already seen, are lined by structures that almost certainly include the fibers of the nucleoskeleton, it is possible that the protein becomes (largely) bound to the nucleoskeleton and is transported to its target site in the solid phase; but it is equally possible a priori that the protein moves in the fluid phase by currents which may be generated by the nucleoskeleton itself.

In short, although the cytoskeleton and nucleoskeleton probably do play parts in karyophilic protein uptake, these parts may be indirect in most cases (Nopp 140 might be exceptional). As with mRNA transport, protein transport is related to the skeletal fibrils in some way, but the mechanisms and certainly the molecular details remain uncertain.

TRANSPORT AND THE MAINTENANCE OF STRUCTURE

This way of expressing the issue implicitly presupposes the cytoskeleton and nucleoskeleton to be permanent, fixed structures that remain essentially unchanged by the migration of the transportant. In the case of the nucleoskeleton at least, this might not be the case.

A general argument favoring the role of the nucleoskeleton in maintaining normal nuclear RNA restriction is that isolated nuclei behave physiologically in respect of mRNA efflux only if they remain unswollen,[62] and morphologically well-formed nuclear matrices[106] cannot be isolated from swollen nuclei. Both cytochalasins and RNAases liberate RNA and at the same time disrupt the reticular fibrillar network of these isolated matrices, suggesting that RNA is a structural component of them;[106] comparable results have been reported by others.[47,50] By implication, RNA may be a structural component of the in situ nucleoskeleton. Generalizing the inference: the migrating transportant might itself be a part of the transporting apparatus, which therefore exists as a complete entity only when it is functioning in the movement of macromolecules. This model contrasts sharply with the view of the nucleoskeleton as a fixed 'railway' or 'cable elevator'[107] that is essentially unaltered by transportant migration.

The new hypothesis is less improbable than it first appears, at least in the case of mRNA precursors. First, we recall that splicing intermediates are actin-associated[63] while spliceosomes are associated with NuMA. There is no known direct connection between these two putative fibrillar components of the nucleoskeleton, so it is feasible that they are joined together mainly by spliceosome-associated RNA. Second, we noted in chapter 4 the evidence linking pore-complex distribution in the nuclear envelope indirectly to normal poly(A)+ RNA transport in yeasts,[108-113] and this becomes more readily understood if pore-complex arrangement is underpinned by a nucleoskeleton, the integrity of which is maintained by mRNA transport. Third, although the dynamics of NuMA fibrils during interphase are not known, the repeated disassembly and reassembly of short F-actin fibrils poses no conceptual problem. Fourth, the mere fact that the nucleoskeleton has long proved elusive and even today has a somewhat debatable in situ character (chapter 3) might suggest its lability and its structural dependence on ephemera such as mRNA migration events. Fifth, the analogy between the intranuclear migrations of karyophilic proteins and messenger precursors discussed in the previous section implies that protein transport might also be relevant to the maintenance of a labile skeleton. The role of the nucleolus in packaging and exporting poly(A)+ RNA[74] may be related to the fact that yeast SRP-1 has been shown to be important in securing pore-complex NUP-1 to the nucleoskeleton and/or nucleolus and in maintaining the integrity of the nucleolus.[114,115] We therefore suggest that both mRNA precursors and (perhaps) karyophilic proteins become integral structural components of a nucleoskeleton during intranuclear migration.

In this context, it is interesting to reconsider the role of NUP-153, the pore-complex component associated with the nuclear basket which is widely supposed to be linked to the nucleoskeleton. NUP-153 has four zinc finger motifs and binds DNA in vitro,[116] but in situ it could be involved alternatively in binding RNA or proteins. If it binds DNA its role could be to maintain open connections with the extrachromosomal network or, perhaps, to facilitate transcription close to the site of translocation. If it binds RNA its role could be related to RNA processing or migration. If it binds proteins then it could be a transducer of conformational changes related to channel opening during transcription. Since several copies are presumably present at each pore-complex it could serve a combination of these functions, securing the correct topological relationship between the chromatin and the nucleoskeleton (keeping the latter on the surface of the former, in contact with active genes) and accepting messengers for translocation at the end of their journey 'through' the nucleoskeleton. The dynamic model we propose does not make these functions mutually exclusive, but rather complementary, and is perhaps supported by the observation that, at least in oocytes, NUP-153 is present in intranuclear fibrils as well as pore-complexes.[69] We might add that our model can be seen as a compromise between the 'fixed' solid-state transport hypothesis and the 'channeled diffusion' hypothesis of Zachar et al,[58] since it envisages the migrating RNA as neither in solution nor as attached to a permanent fibrillar system.

An obvious objection to this 'interactive perspective' on mRNA transport and a fortiori to 'interactive' models of nucleocytoplasmic transport concerns the generation of movement. At first sight it is difficult to see how, for instance, a messenger strung between a NuMA and an actin fibril like a mountaineer in an overly wide rock chimney can migrate to a different but similar location. Two possible mechanisms suggest themselves. First, if some other step in the overall transport is made (largely) vectorial at the expense of ATP, then movement through the rest of the system could be generated by mass action. As we saw in chapter 4, translocation through the pore-complex is usually energy-dependent. Second, short-range thermal perturbations will ensure rapid dissociation and reassociation events, especially since the relevant K_d values are high, and a mechanism such as that outlined in Figure 5.3 might operate. This second mechanism is not incompatible with our argument in chapter 2. Diffusion cannot account for long-range macromolecule movements in cells, but Brownian motion is inevitable within distances of a few tens of nanometers (the dimensions of the supposed 'microtrabecular lattice'); without it, binding and dissociation events including those involved in enzyme catalysis could not occur.

We hesitate to extrapolate our model to the cytoskeleton. Nevertheless, some simple experiments might be worth trying. The assembly dynamics of F-actin and other fibril types in vitro might be altered in the presence of karyophilic proteins and LSBs such as importins, and the implications of positive results could be interesting.

Because the model is general it does not allow explicitly for polarized distributions of particular mRNAs in the cytoplasm as observed in the pair-rule transcripts in early *Drosophila* embryos,[117] nor does it obviously accommodate the cytoskeletal partitioning of some messengers as described earlier in this chapter. The polarity might result at least in part from the utilization of only a limited number of pore-complexes for translocating the relevant messengers, perhaps because they are transcribed and processed close to the nuclear surface at the appropriate point. This recalls the gene gating hypothesis of Blobel[118] which is by no means incompatible with the model we propose. However, most messengers probably leave the pore-complexes unselectively, giving observational data of the type reported by Zachar et al,[58] originating from transcription and processing sites more remote from the pore-complex.

THE FUNCTIONS OF RAN AND RCC1

In chapter 4 we reasoned that the nuclear GTPase cycle probably had a role in transport immediately outside the pore-complex and/or within the nucleus rather than in transcription through the pore-complex per se, perhaps removing LSBs from karyophilic proteins after, and from mRNAs before, transit of the pore-complex. In the light of the model we have proposed for the involvement of the nucleoskeleton in transport we need to reconsider and develop this argument. We recall that the guanine nucleotide exchange factor RCC1 may be part of a higher order structure,[119] perhaps chromosome-associated,[120] and that the nuclear Ran binding proteins are insoluble and resist extraction with Triton and high ionic strength media.[121] There is now abundant evidence that Ran and RCC1 are involved in processes other than nucleocytoplasmic transport, and nucleoskeleton dynamics might be common to these processes.

Loss of function of Ran and/or RCC1 results in errors in the termination of transcription leading to the production of messengers with extended 3' ends to defective processing of rRNA and tRNA and in yeasts to fragmentation of the nucleolus.[122-125] Impaired transcription and nucleolar fragmentation suggest perturbation of nucleoskeletal organization which would, according to our hypothesis, be tantamount to inhibition of pre-mRNA migration within the nucleus. The implication is that normal rRNA and tRNA processing also require a normal nucleoskeleton. Another traditionally nucleoskeleton-related function[47,54] that is impaired if Ran or RCC1 is defective is DNA replication, though DNA polymerase activity is not affected.[126,127] Again, a nucleoskeleton abnormality that prevents functional DNA binding is suggested here. Moreover, defects of this kind lead to premature mitotic chromosome condensation even if DNA synthesis is inhibited, so long as p34[cdc2] is active and protein synthesis occurs; in some cases chromatin fragmentation and the formation of micronuclei result.[126,127,129-133] There is increasing evidence for the involvement of RCC1 and Ran in a wide range of nuclear structures and functions.[134-137] We recall the evidence implying that the nucleoskeleton is extended over the surfaces of decondensed chromatin; perhaps this association is necessary to maintain the decondensed state and therefore nucleoskeletal defects cause premature condensation if other factors necessary for the initiation of mitosis are present in the cell.

In summary, it is quite possible that the nuclear GTPase cycle components are essential for the normal functional organization of the nucleoskeleton, and inhibition of them therefore results in inhibition of DNA replication, RNA processing, intranuclear pre-mRNA transport and protein import as well as leading to nucleolar fragmentation and premature chromosome condensation. It therefore becomes appropriate to inquire which nucleoskeletal component is affected by Ran. In chapter 4 we noted that one of the Ran binding proteins in the nucleus might be NuMA, but none of these proteins corresponds in M_r or other biochemical properties to actin. There is also evidence that microinjection of antibodies against NuMA causes nuclear fragmentation of a type morphologically similar to that associated with GTPase cycle inhibition[130,138] and overexpression of NuMA in yeasts partially rescues GTPase cycle component mutations. This evidence is not definitive and interactions between NuMA, Ran and RCC1 would be worth examining in detail. Nevertheless, it does suggest that NuMA is the principal target of Ran activity and that GTPase impairment initiates a succession of functional abnormalities of the kind suggested in Fig. 5.4. On the other hand, at least some isoforms of NuMA fail to bind Ran-GTP in vitro.[139]

However, the system has other subtleties beyond the scope of this scheme. Proliferating cell nuclei are more active than quiescent ones in the uptake of karyophilic substrates, not surprisingly in view of the greater material demands on the nucleus when the cell is cycling.[140] The Ran-RCC1 associated factor pp15 was shown to be necessary for nuclear protein uptake in cells that were probably quiescent.[121] When proliferating HeLa cells were permeabilized and nuclear protein import was reconstituted by addition of Ran and importins the results were heterogeneous, some nuclei showing very marked fluorescence and, often, concomitant chromatin decondensation, and the rest showing little or no fluorescence.[141] Görlich et al[141] suggest that this heterogeneity might have resulted

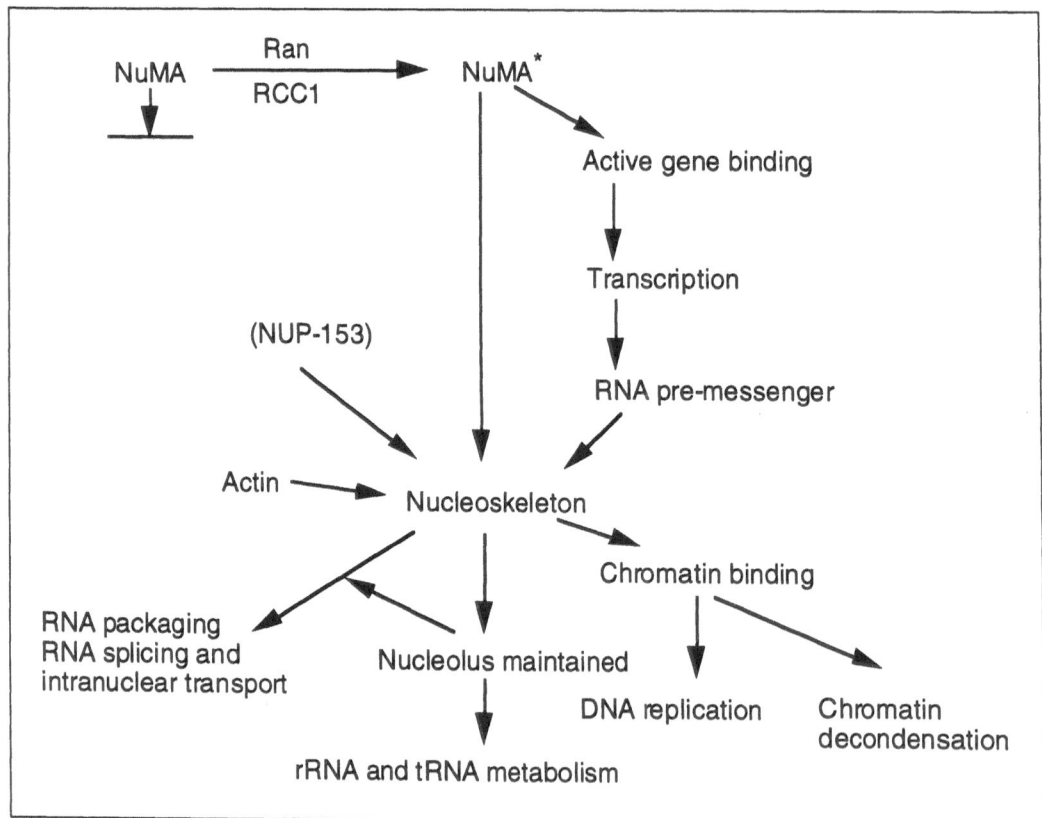

Fig. 5.4. *Widespread consequences of Ran-RCC1 actions. Assuming that the nuclear GTPase cycle components act principally by modifying NuMA, and that NuMA is a nucleoskeletal component capable of binding active genes, organizing chromatin, and (in association with actin and NUP-153) of packaging, modifying and transporting RNA, the widespread and varied effects of Ran-RCC1 in the nucleus can be explained by the sequences of events illustrated.*

from the absence of pp15, which is perhaps essential if the cells are in G_0 or G_1 but not otherwise. Given that nucleoskeletal organization is presumably altered during S phase (since its fibrils are in immediate contact with the replicating or replicated DNA), it may be that an additional factor such as pp15 is needed for transport only when the organization is characteristic of nondividing cells. This speculation cannot be tested until the nucleoskeleton is better characterized, and particularly until the molecular details of the interactions amongst NuMA, F-actin, premessengers and karyophilic proteins have been explored.

NUCLEOLIN: IS IT IDENTICAL WITH P110?

The major nucleolar structural protein nucleolin might play a central role in nuclear RNA and protein transport. Its levels increase in response to transcriptional activity[142] and it can behave as a helicase.[143] It can also act as LSB for imported proteins.[144] Moreover, it has striking resemblances to P110, suggesting homology, not only in M_r, but also in phosphorylation states,[145] and in proteinase lability, which might include self-cleaving.[146,147] Nucleolin and hsp70, a known cofactor in RNA export, are induced together in regenerating liver,[148] and nucleolin along with the C group ribonucleosome core proteins binds specifically to the 3' untranslated regions of certain pre-mRNAs.[149] Therefore, the role of the nucleolus in nucleocytoplasmic processes might be attributed to the fact that nucleolin is a major component.

OVERVIEW

In the past, two perspectives of nucleocytoplasmic transport (particularly mRNA transport) have been adopted. The first and more widely adopted supposed the pore-complex to be the principal or even the sole cellular structure of relevance; we cast doubt on it in the early chapters of this book and found in chapter 4 that it is inadequate to explain the data, including data from experiments designed under the presumption that it is valid. The second, the solid-state transport perspective, held that transport is a continuous process involving nucleoskeleton and cytoskeleton at least as much as the pore-complex, but was often taken to imply that mRNA migration involved a continuous fibrillar system extending throughout nucleoplasm and cytoplasm via the pore-complexes (for a recent evaluation see ref. 101). The difficulty with the solid-state perspective has been constructing a more precise model compatible with all the data.[101]

The model we have evolved during the course of this chapter acknowledges the sense of 'transport' implied in the solid-state perspective but differs from the traditional solid-state idea in two important particulars. It pictures the nucleoskeleton as an intact integral structure only when material is being transported, thus avoiding some of the difficulties of most solid-state models,[100] and accommodating evidence that seems initially to point to events such as 'diffusion' during intranuclear migration.[58] Also, it suggests that messengers do not bind specifically to the cytoskeleton immediately on exiting the pore-complex, but partition amongst at least two classes of binding sites from a 'free' pool.[9,40] We shall develop this model and its implications in the remaining two chapters. In chapter 6 we shall examine its performance with regard to information about nucleocytoplasmic transport pathologies and the possibilities for regulating the distributions of transportants between the two major cell compartments. In chapter 7 we shall consider its possible generality.

References

1. Rao MVD, Prescott DM. Return of RNA into the nucleus after mitosis. J Cell Biol 1967; 35:109a.
2. Drummond DR, McGrae MA, Colman A. Stability and movement of mRNAs and their encoded proteins in *Xenopus* oocytes. J Cell Biol 1985; 100:1148-1156.
3. Cervera M, Dreyfuss G, Penman S. Messenger RNA is translated when associated with the cytoskeletal framework in normal and vesicular stomatitis virus infected HeLa cells. Cell 1981; 23:113-120.
4. Van Venrooij WJ, Sillekens PTG, van Eekelen CAG et al. On the association of mRNA with the cytoskeleton in uninfected and adenovirus infected human KB cells. Exp Cell Res 1981; 135:79-92.
5. Jones NL, Kirkpatrick BA. The effects of human cytomegalovirus infection on cytoskeleton associated polysomes. Eur J Cell Biol 1987; 46:565-575.
6. Sundell CL, Singer RH. Actin mRNA localizes in the absence of protein synthesis. J Cell Biol 1990; 111:2397-2403.
7. Sundell CL, Singer RH. Requirement of microfilaments in sorting of actin messenger RNA. Science 1991; 253:1275-1277.
8. Walter WB, Gilmore R, Blobel G. Protein translocation across the endoplasmic reticulum. Cell 1984; 38:5-8.
9. Hesketh JE, Pryme IF. Interaction between mRNA, ribosomes and the cytoskeleton. Biochem J 1991; 277:1-10.

10. Lawrence JB, Singer RH. Intracellular localization of messenger RNAs for cytoskeletal proteins. Cell 1986; 45:407-415.
11. Cheng H, Bjerknes M. Asymmetric distribution of actin mRNA and cytoskeletal pattern generation in polarized epithelial cells. J Mol Biol 1989; 210:541-549.
12. Hoock TC, Newcomb PM, Herman IM. Beta-actin and its mRNA are localized at the plasma membrane and the regions of moving cytoplasm during the cellular response to injury. J Cell Biol 1991; 112:653-664.
13. Garner CC, Tucker RP, Matus A. Selective localization of messenger RNA for cytoskeletal protein MAP2 in dendrites. Nature 1988; 336:674-677.
14. Kleiman R, Banker G, Steward O. Differential subcellular localization of particular mRNAs in hippocampal neurons in culture. Neuron 1990; 5:821-830.
15. Kleiman R, Banker, G, Steward O. Development of subcellular mRNA compartmentation in hippocampal neurons in culture. J Neurosci 1994; 14:1130-1140.
16. Pomeray ME, Lawrence JB, Singer RH et al. Distribution of myosin heavy chain mRNA in embryonic muscle tissue visualized by ultrastructural *in situ* hybridization. Dev Biol 1991; 143:58-67.
17. Singer RH. The cytoskeleton and mRNA localization. Curr Opin Cell Biol 1992; 4:15-19.
18. Vedeler A, Pryme IF, Hesketh JE. The characterization of free, cytoskeletal and membrane-bound polysomes in Krebs II ascites and 3T3 cells. Mol Cell Biochem 1991; 100:183-193.
19. Ding D, Lipshitz HD. Localized RNAs and their functions. Bioessays 1993; 15:651-658.
20. Davenport R. Transport of ribosomal RNA into the oocytes of the milkweed bug, *Oncopeltus fasciatus*. J Insect Physiol 1976; 22:925-926.
21. Pokrywka NJ, Stephenson EC. Microtubules mediate the localization of bicoid RNA during *Drosophila* oogenesis. Development 1991; 113:55-66.
22. Lane ME, Kalderon D. RNA localization along the anteroposterior axis of the *Drosophila* oocyte requires PKA-mediated signal transduction to direct normal microtubule organization. Genes Dev 1994; 8:2986-2995.
23. Ainger K, Avossa D, Morgan F et al. Transport and localization of exogenous myelin basic protein mRNA microinjected into oligodendrocytes. J Cell Biol 1993; 123:431-441.
24. MacDonald PM, Struhl G. Cis-acting sequences responsible for anterior localization of bicoid messenger-RNA in *Drosophila* embryos. Nature 1988; 336:595-598.
25. Macdonald PM, Kerr K, Smith JL et al. RNA regulatory element BLE1 directs the early steps of bicoid mRNA localization. Development 1993; 118:1233-1243.

26. Ferrandon D, Elphick L, Nusslein-Volhard C et al. Staufen protein associates with the 3'UTR of bicoid mRNA to form particles that move in a microtubule-dependent manner. Cell 1994; 79:1221-1232.

27. Yang F, Demma M, Warren V et al. Identification of an actin-binding protein from *Dyctostelium* as elongation factor 1. Nature 1990; 347:494-496.

28. Rhoads, RE. Cap recognition and the entry of mRNA into the protein synthesis initiation cycle. Trends Biochem Sci 1988; 13:52-56.

29. Pause A, Methot N, Svitken Y et al. Dominant negative mutants of mammalian translation initiation factor eIF-4A define a critical role for eIF-4F in cap-dependent and cap-independent initiation of translation. EMBO J 1994; 13:1205-1215.

30. Schröder H-C, Diehl-Seifert B, Rottmann M et al. Functional dissection of nuclear envelope mRNA translocation system: effects of phorbol ester and a monoclonal antibody recognizing cytoskeletal structures. Arch Biochem Biophys 1988; 261:394-404.

31. Schröder H-C, Rottmann M, Bachmann M et al. Proteins from rat liver cytosol which stimulate mRNA transport: purification and interactions with the nuclear envelope mRNA translocation system. Eur J Biochem 1986; 159:51-59.

32. Müller WEG, Agutter PS, Schröder HC. Transport of mRNA into the cytoplasm. In: Maceiera-Coelho A, ed. Molecular Basis of Aging. Boca Raton, New York, London and Tokyo: CRC press, 1995:353-388.

33. Yisraeli JK, Sokol S, Melton DA. A two step model for the localization of maternal mRNA in *Xenopus* oocytes. Development 1990; 108:289-298.

34. Jeffery WR. Localized mRNA and the egg cytoskeleton. Int Rev Cytol 1989; 119:151-193.

35. Lenk R, Ransom L, Kaufmann Y et al. A cytoskeletal structure with associated polyribosomes obtained from HeLa cells. Cell 1977; 10:67-78.

36. Scherrer K. Prosomes, subcomplexes of untranslated messenger RNP. Mol Biol Rep 1990; 14:1-9.

37. Georgatos SD, Blobel G. Lamin B constitutes an intermediate filament attachment site at the nuclear envelope. J Cell Biol 1987; 105:117-125.

38. Moon RT, Nicosia RF, Olsen C et al. The cytoskeletal framework of sea urchin eggs and embryos. Developmental changes in the association of messenger RNA. Dev Biol 1983; 95:447-458.

39. Vedeler A, Pryme IF, Hesketh JE. Insulin and step-up conditions cause a redistribution of polysomes among free, cytoskeletal-bound and membrane-bound fractions in Krebs II ascites cells. Cell Biol Int Rep 1990; 14:211-218.

40. Zambetti G, Fey EG, Penman S et al. Multiple types of mRNA cytoskeleton interactions. J Cell Biochem 1990; 44; 177-187.
41. Darnell JE. Variety in the level of gene control in eukaryotic cells. Nature 1982; 297:365-371.
42. Nakayasu H, Ueda K. Small nuclear RNP complex anchors on the actin filaments in bovine lymphocyte nuclear matrix. Cell Struct Funct 1984; 9:317-326.
43. Schindler M, Jiang L-W. Nuclear actin and myosin as control elements in nucleocytoplasmic transport. J Cell Biol 1986; 102: 859-862.
44. Bladon T, Brasch KR, Brown DL et al. Changes in structure and composition of bovine lymphocyte nuclear matrix during concanavalin-A-stimulated mitogenesis. Biochem Cell Biol 1988; 66:40-53.
45. Long BM, Wang C-Y, Pogo AO. Isolation and characterization of the nuclear matrix in Friend erythroleukemia cells: chromatin and HnRNA interactions with the nuclear matrix. Cell 1979; 18: 1079-1090.
46. Agutter PS, Birchall K. Functional differences between mammalian nuclear matrix and pore-lamina preparations. Exp Cell Res 1979; 124:453-460.
47. Berezney R. Dynamics of the nuclear protein matrix. In: Busch H, ed. The Cell Nucleus. Vol. 7. New York and London: Academic Press 1980:413-455.
48. Van Eekelen CAG, van Venrooij WJ. HnRNA and its attachment to a nuclear matrix. J Cell Biol 1981; 88:554-563.
49. Habets WJ, Berden JHM, Hoch SO et al. Further characterization and subcellular localization of Sm and U1 ribonucleoprotein antigens. Eur J Immunol 1985; 15:992-997.
50. Fey EG, Krochmalnic G, Penman S. The nonchromatin substructures of the nucleus: the RNP-containing and RNP-depleted matrices analyzed by sequential fractionation and resinless electron microscopy. J Cell Biol 1986; 102:1654-1665.
51. Smith HC, Ocho RL, Fernandez EA et al. Macromolecular domains containing nuclear P-107 and UsnRNP protein P-28: further evidence for an *in situ* nuclear matrix. Mol Cell Biochem 1986; 70:151-168.
52. Verheijen R, Kuijpers H, Voojs P et al. Distribution of the 70K U1 RNA associated protein during interphase and mitosis: correlation with other U RNP particles and proteins of the nuclear matrix. J Cell Sci 1986; 86:173-190.
53. Feldherr CM. Ribosomal RNA synthesis and transport following disruption of the nuclear envelope. Cell Tissue Res 1980; 205:157-162.
54. Jackson DA, Cook PR. Transcription occurs at a nuclear cage. EMBO J 1985; 4:919-926.
55. Lawrence JB, Singer RH, Marselle LM. Highly localized tracks of

specific transcripts within interphase nuclei visualised by *in situ* hybridization. Cell 1989; 57:493-502.

56. Huang S, Spector DL. Nascent pre-mRNA transcripts are associated with nuclear regions enriched in splicing factors. Genes Dev 1991; 5:2288-2302.

57. Xing Y, Lawrence JB. Preservation of specific RNA distribution within the chromatin-depleted nuclear substructure demonstrated by *in situ* hybridization coupled with biochemical fractionation. J Cell Biol 1991; 112:1055-1064.

58. Zachar Z, Kramer J, Mims IP et al. Evidence for channelled diffusion of pre-mRNAs during nuclear transport in metoazoans. J Cell Biol 1993; 121:729-742.

59. Rosbash M, Singer RH. RNA travel: Tracks from DNA to cytoplasm. Cell 1993; 75:399-401.

60. Wang J, Cao LG, Wang Y-L et al. Localization of pre-messenger RNA at discrete nuclear sites. Proc Natl Acad Sci USA 1991; 88:7391-7395.

61. Barja F, Turian G. Cytochalasin B-sensitive actin-mediated nuclear RNA export in germinating conidia of *Neurospora crassa*. Cell Biol Internat 1994; 18:903-906.

62. Agutter PS. An assessment of some methodological criticisms of RNA efflux studies using isolated nuclei. Biochem J 1983; 214:915-921.

63. Schröder HC, Trölltsch D, Wenger R et al. Cytochalasin-B selectively releases ovalbumin messenger-RNA precursors but not the mature ovalbumin messenger-RNA from hen oviduct nuclear matrix. Eur J Biochem 1987; 167:239-245.

64. Sahlas DJ, Milankov K, Park PC et al. Distribution of snRNPs, splicing factor SC-35 and actin in interphase nuclei: Immunocytochemical evidence for differential distribution during changes in functional states. J Cell Sci 1993; 259:1330-1335.

65. Carter KC, Bowman D, Carrington W et al. A three-dimensional view of precursor messenger RNA metabolism within the mammalian nucleus. Science 1993; 259:1330-1335.

66. Meier T, Blobel G. Nopp 140 shuttles on tracks between nucleolus and cytoplasm. Cell 1992; 70:127-138.

67. Murti KG, Brown PS, Ratner L et al. Highly localized tracks of human immunodeficiency virus type 1 Nef in the nucleus of cells of a human CD4+ T-cell line. Proc Natl Acad Sci USA 1993; 90:11895-11899.

68. Miller M, Park MK, Hanover JA. Nuclear pore complex: Structure, function, and regulation. Physiol Rev 1991; 71:909-949.

69. Cordes VC, Reidenback S, Kohler A et al. Intranuclear filaments containing a nuclear pore complex protein. J Cell Biol 1993; 123:1333-1344.

70. Hozak P, Sasseville AMJ, Raymond Y et al. Lamin proteins form

an internal nucleoskeleton as well as a peripheral lamina in human cells. J Cell Sci 1995; 108:635-644.

71. Wansink DG, Manders EEM, Van Der Kraan I et al. RNA polymerase II transcription is concentrated outside replication domains throughout S-phase. J Cell Sci 1994; 107:1449-1456.

72. Lamond AI, CarmoFonseca M. The coiled body. Trends Cell Biol 1993; 3:198-204.

73. Bohmann K, Ferriera JA, Lamond AI. Mutational analysis of p80 coilin indicates a functional interaction between coiled bodies and the nucleolus. J Cell Biol 1995; 131:817-831.

74. Schneiter R, Kadowaki T, Tartakoff AM. mRNA transport in yeast: Time to reinvestigate the functions of the nucleolus. Mol Biol Cell 1995; 6:357-370.

75. Sibon OCM, Cremers FFM, Boonstra J et al. Localisation of EGF-receptor mRNA in the nucleus of A431 cells by light microscopy. Cell Biol Internat 1993; 17:1-11.

76. Lindersen BK, Pettijohn DE. Human specific nuclear protein that associates with the polar region of the mitotic apparatus: distribution in a human/hamster hybrid cell. Cell 1980; 22:489-499.

77. Zeng C, He D, Berget SM et al. Nuclear-mitotic apparatus protein: A structural protein interface between the nucleoskeleton and RNA splicing. Proc Natl Acad Sci USA 1994; 91:1505-1509.

78. Zeitlin S, Parent A, Silverstein S et al. Pre-mRNA splicing and the nuclear matrix. Mol Cell Biol 1987; 7:111-120.

79. Gruss P, Lai C-J, Dhar R et al. Splicing as a requirement for biogenesis of functional 16S mRNA of SV$_{40}$. Proc Natl Acad Sci USA 1979; 76:4317-4321.

80. De la Peña P, Zasloff M. Enhancement of mRNA nuclear transport by promotor elements. Cell 1987; 50:613-619.

81. Hamm J, Mattaj IW. Monomethylated cap structures facilitate RNA export from the nucleus. Cell 1990; 63:109-118.

82. Izzauralde E, Lewis J, Gamberi C et al. A cap binding protein complex mediating U snRNA export. Nature 1995; 376:709-712.

83. Mehlin H, Daneholt B, Skoglund U. Structural interaction between the nuclear pore complex and a specific translocating RNP particle. J Cell Biol 1995; 129:1205-1216.

84. Plumpton M, McGarvey M, Beggs JD. A dominant negative mutation in the conserved RNA helicase motif 'SAT' causes splicing factor PRP2 to stall in spliceosomes. EMBO J 1994; 13:879-887.

85. Tieglekamp S, McGarvey M, Plumpton M et al. The splicing factor PRP2, a putative RNA helicase, interacts directly with pre-mRNA. EMBO J 1994; 13:888-897.

86. Le Stourgeon WM. The occurrence of contractile proteins in nuclei and their possible functions. In: Busch H, ed. The Cell Nucleus. Vol 6. New York: Academic Press, 1978:305-326.

87. Berrios M, Fisher PA. A myosin heavy chain-like polypeptide is

associated with the nuclear envelope in higher eukaryotic cells. J Cell Biol 1986; 103:711-724.

88. Berrios M, Osherhoff N, Fisher PA. *In situ* localization of DNA topoisomerase II, a major polypeptide component of the *Drosophila* nuclear matrix. Proc Natl Acad Sci USA 1985; 82:4142-4146.

89. Schröder H-C, Trölltsch D, Friese U et al. Mature messenger RNA is selectively released from the nuclear matrix by an ATP-deoxy-ATP-dependent mechanism sensitive to topoisomerase inhibitors. J Biol Chem 1987; 262:8917-8925.

90. Watt PM, Hickson ID. Structure and function of type II DNA topoisomerases. Biochem J 1994:303:681-691.

91. Vassetzky YS, Dang Q, Benedetti P et al. Topoisomerase II forms multimers in vitro: Effects of metals, beta-glycerophosphate, and phosphorylation of its C-terminal domain. Mol Cell Biol 1994; 14:6962-6974.

92. Haselbeck RC, Greer CL. Minimum intron requirements for tRNA splicing and nuclear transport in *Xenopus* oocytes. Biochemistry 1993; 32:8575-8581.

93. Paine PL. Diffusive and nondiffusive proteins in vivo. J Cell Biol 1984; 99:188s-195s.

94. Feldherr CM, Ogburn JA. Mechanisms for the selection of nuclear polypeptides in *Xenopus* oocytes. II: Two-dimensional gel analysis. J Cell Biol 1980; 87:589-593.

95. Paine PL. Nuclear protein accumulation by facilitated transport and intranuclear binding. Trends Cell Biol 1993; 3:325-329.

96. Smith AE. The nuclear location signal. Proc Roy Soc B 1985; 226:43-58.

97. Yamasaki L, Kanda P, Lanford RE. Identification of four nuclear transport signal-binding proteins that interact with diverse transport signals. Mol Cell Biol 1989; 9:3028-3036.

98. Breeuwer M, Goldfarb DS. Facilitated nuclear transport of histone H1 and other small nucleophilic proteins. Cell 1990; 60:999-1008.

99. Stochaj U, Silver PA. A conserved phosphoprotein that specifically binds nuclear localization sequences is involved in nuclear import. J Cell Biol 1992; 117:473-482.

100. Ambron RT, Schmied R, Huang CC et al. A signal sequence mediates the retrograde transport of proteins from the axon periphery to the cell body and then into the nucleus. J Neurosci 1992; 12:2813-2818.

101. Agutter PS. Models for solid-state transport: messenger RNA movement from nucleus to cytoplasm. Cell Biol Internat 1994; 18:849-858.

102. Nirmala PB, Varman Thampan R. A 55KDa protein (p55) of the goat uterus mediates nuclear transport of the oestrogen receptor. I. Purification and characterization. Arch Biochem Biophys 1995; 319:551-561.

103. Vancurova I, Jochova J, Lou W et al. An NLS is sufficient to engage facilitated translocation by the nuclear pore complex and subsequent intranuclear binding. Biochem Biophys Res Commun 1994; 205:529-536.

104. Schmidt-Zachmann MS, Dargemont C, Kühn LC et al. Nuclear export of proteins: the role of nuclear retention. Cell 1993; 74:493-504.

105. Allison LA, North MT, Murdoch KJ et al. Structural requirements of 5SrRNA for nuclear transport 7S ribonucleoprotein particle assembly, and 60S ribosomal subunit assembly in *Xenopus* oocytes. Mol Cell Biol 1993; 13:6819-6831.

106. Comerford SA, Agutter PS, McLennan AG. Nuclear matrices. In: MacGillivray AJ, Birnie GD, eds. Nuclear Stuctures: Their Isolation and Characterisation. London: Butterworth, 1986:1-13.

107. Maul GG, ed. The Nuclear Envelope and the Nuclear Matrix. New York: Alan R Liss, 1982:1-11.

108. Doye V, Wepf R, Hurt EC. A novel nuclear pore protein Nup133p with distinct roles in poly(A)+ RNA transport and nuclear pore distribution. EMBO J 1994; 13:6062-6075.

109. Li O, Heath CV, Amberg DC et al. Mutation or deletion of the *Saccharomyces cerevisiae* RAT3/NUP133 gene causes temperature-dependent nuclear accumulation of poly(A)+ RNA and constitutive clustering of nuclear pore complexes. Mol Biol Cell 1995; 6:401-417.

110. Gorsch LC, Dockendorff TC, Cole CN. A conditional allele of the novel repeat-containing yeast nucleoporin RAT7/NUP159 causes both rapid cessation of mRNA export and reversible clustering of nuclear pore complexes. J Cell Biol 1995; 129:939-955.

111. Fabre E, Boelens WC, Wimmer C et al. Nup145p is required for nuclear export of mRNA and binds homopolymeric RNA in vitro via a novel conserved motif. Cell 1994; 78:275-289.

112. Wente SR, Blobel G. NUP145 encodes a novel yeast glycine-leucine-phenylalanine-glycine (GLFG) nucleoporin required for nuclear envelope structure. J Cell Biol 1994; 125:955-959.

113. Pemberton LF, Rout MP, Blobel G. Disruption of the nucleoporin gene NUP133 results in clustering of nuclear pore complexes. Proc Natl Acad Sci USA 1995; 92:1187-1191.

114. Belanger KD, Kenna MA, We S et al. Genetic and physical interactions between $SRP1_p$ and nuclear pore-complex proteins Nup1p and Nup2p. J Cell Biol 1994; 126:619-630.

115. Yano R, Oakes ML, Tabb MM et al. Yeast $SRP1_p$ has homology to armadillo/plakoglobin/β-catenin and participates in apparently multiple-nuclear functions including the maintenance of nucleolar structure. Proc Natl Acad Sci USA 1994; 91:6880-6884.

116. Sukegawa J, Blobel G. A nuclear pore complex protein that contains zinc finger motifs, binds DNA, and faces the nucleoplasm.

Cell 1993; 72:29-38.

117. Davis I, Ish-Horowicz D. Apical localization of pair-rule transcripts requires 3' sequences and limits protein diffusion in the *Drosophila* blastoderm embryo. Cell 1991; 67:927-940.

118. Blobel G. Gene gating; an hypothesis. Proc Natl Acad Sci USA 1985; 82:8527-8529.

119. Ren M, Drivas G, d'Eustachio P et al. Ran/TC4: A small nuclear GTP-binding protein that regulates DNA synthesis. J Cell Biol 1993; 120:313-323.

120. Melchior F, Paschal B, Evans E et al. Inhibition of nuclear protein import by nonhydrolyzable analogues of GTP and identification of the small GTPase Ran/TC4 as an essential transport factor. J Cell Biol 1993; 123:1649-1659.

121. Moore MS, Blobel G. Purification of a ran-interacting protein that is required for protein import into the nucleus. Proc Natl Acad Sci USA 1994; 91:10212-10216.

122. Forrester W, Stutz F, Rosbash M et al. Defects in mRNA 3'-end formation, transcription initiation, and mRNA transport associated with the yeast mutation prp20: Possible coupling of mRNA processing and chromatin structure. Genes Dev 1992; 6:1914-1926.

123. Oakes M, Nagi M, Clark M et al. Structural alterations of the nucleolus in mutants of Saccharomyces cerevisiae defective in RNA polymerase I. Mol Cell Biol 1993; 13:2441-2455.

124. Kadowaki, T, Goldfarb D, Spitz L et al. Regulation of RNA processing and transport by a nuclear guanine nucleotide release protein and members of the Ras superfamily. EMBO J 1993; 12: 2929-2937.

125. Kadowaki T, Hitomi M, Chen S et al. Mutations in nucleolar proteins lead to nucleolar accumulation of PolyA+ RNA in Saccharomyces cerevisiae. Mol Biol Cell 1995; 6:1103-1110.

126. Dasso M, Nishitani H, Kornbluth S et al. RCC1, A regulator of mitosis, is essential for DNA-replication. Mol Cell Biol 1992; 12:3337-3345.

127. Kornbluth S, Dasso M, Newport J. Evidence for a dual role for TC4 protein in regulating nuclear structure and cell cycle progression. J Cell Biol 1994; 125:705-719.

128. Sazer S, Nurse P. A fission yeast RCC1-related protein is required for the mitosis to interphase transition. EMBO J 1994; 13:606-615.

129. Nishimoto T, Eilen E, Basilico C. Premature chromosome condensation in a ts DNA-mutant of BHK cells. Cell 1978; 15: 475-483.

130. Nishitani H, Ohtsubo M, Yamashita K et al. Loss of RCC1, a nuclear DNA-binding protein, uncouples the completion of DNA replication from the activation of cdc2 protein kinase and mitosis. EMBO J 1991; 10:1555-1564.

131. Matsumoto T, Beach D. Premature initiation of mitosis in yeast

lacking RCC1 or an interacting GTPase. Cell 1991; 66:347-360.

132. Seino H, Hisamoto N, Uzawa S et al. DNA-binding domain of RCC1 protein is not essential for coupling mitosis with DNA replication. J Cell Sci 1992; 102:393-400.

133. Dasso M. RCC1 in the cell cycle: The regulator of chromosome condensation takes on new roles. Trends Biochem Sci 1993; 18:96-101.

134. Dasso M, Seki T, Azuma Y et al. A mutant form of the Ran/TC4 protein disrupts nuclear function in *Xenopus laevis* egg extracts by inhibiting the RCC1 protein, a regulator of chromosome condensation. EMBO J 1994; 13:5732-5744.

135. Demeter J, Morphew M, Sazer S. A mutation in the RCC1-related protein pim1 results in nuclear envelope fragmentation in fusion yeast. Proc Natl Acad Sci USA 1995; 92:1436-1440.

136. Cheng Y, Dahlberg JE, Land E. Diverse effects of the guanine nucleotide exchange factor RCC1 on RNA transport. Science 1995; 267:1807-1810.

137. Kiss T, Filipowicz W. Small nucleolar RNAs encoded by introns of the human cell cycle regulatory gene RCC1. EMBO J 1993; 12:2913-2920.

138. Compton DA, Cleveland DW. NuMA is required for the proper completion of mitosis. J Cell Biol 1993; 120:947-957.

139. Kempf T, Bischoff FR, Kalies I et al. Isolation of human NuMA protein. FEBS Lett 1994; 354:307-310.

140. Feldherr CM, Akin D. Regulation of nuclear transport in proliferating and quiescent cells. Exp Cell Res 1993; 205:179-186.

141. Görlich D, Kostka S, Kraft R et al. Two different subunits of importin cooperate to recognize nuclear localization signals and bind them to the nuclear envelope. Curr Biol 1995; 5:383-392.

142. Derenzini M, Sirri V, Pessin A et al. Quantitative changes of the two major Ag-NOR proteins, nucleolin and protein B23, related to stimulation of rDNA transcription. Exp Cell Res 1995; 219:276-282.

143. Tuteja N, Huang NW, Skopac D et al. Human DNA helicase IV is nucleolin, an RNA helicase modulated by phosphorylation. Gene 1995; 160:143-148.

144. Xu ZX, Shan XY, Lapeyre B et al. The amino terminus of mammalian nucleolin specifically recognizes SV_{40} T-antigen type nuclear localization sequences. Eur J Cell Biol 1993; 62:13-21.

145. Messmer B, Dreyer C. Requirements for nuclear translocation and nucleolar accumulation of nucleolin of *Xenopus laevis*. Eur J Cell Biol 1993; 61:369-282.

146. Suzuku T, Suzuku N, Hosoya T. Limited proteolysis of rat liver nucleolin by endogenous proteases-effects of polyamines and histones. Biochem J 1993; 289:109-115.

147. Fang SH, Yeh NH. The self-cleaving activity of nucleolin deter-

mines its molecular dynamics in relation to cell proliferation. Exp
Cell Res 1993; 208:48-53.

148. Konishi T, Karasaki Y, Nomoto M. Induction of heat shock protein 70 and nucleolin and their intracellular distribution during
early stage of liver regeneration. J Biochem 1995; 117:1170-1177.

149. Zaidi SHE, Malter JS. Nucleolin and heterogenous nuclear ribonucleoprotein C proteins specifically interact with the 3' untranslated
region of amyloid precursor mRNA. J Biol Chem 1995; 270:
17292-17298.

CHAPTER 6

REGULATION AND PATHOLOGIES OF NUCLEOCYTOPLASMIC TRANSPORT

INTRODUCTION

Nucleocytoplasmic traffic is vital to the cell, in an obviously literal sense. It is no less vital that the traffic be precisely controlled and choreographed. Ensuring nuclear export of specific messengers and import of specific transcription components at the right but not the wrong times in the cell cycle is an intuitive precondition of viability. Thus, the regulation of nucleocytoplasmic exchange processes is as much a part of 'meaning' in the sense of this book's title as the mechanisms themselves.

We shall begin the chapter by surveying our so-far limited knowledge of the control of protein import from cytoplasm to nucleus, and then examine the ways in which the nucleus might respond to extracellular signals. Under these headings we shall reflect briefly on the relationship between nucleocytoplasmic transport and the control of the cell cycle. This brevity is dictated by exiguities of information rather than the importance of the topic, which surely looms large in cell biology, but at least some specific examples can be given. After this, our emphasis will be on mRNA rather than protein transport. There are two fundamental difficulties with this topic. First, although certain specific effectors do appear to modulate mRNA transport, the biological significance of their effects and the mechanisms involved are somewhat mysterious. Second, most of what has been written about the 'regulation' of mRNA transport concerns global, nonspecific changes including those associated with viral infection, carcinogen treatment and aging; we survey this literature here because an adequate model of transport must be able to accommodate the data, but these are quite different in kind from the data on protein transport regulation which are to hand. Throughout the chapter, a now-familiar theme will take on new emphasis; the recognition and articulation of problems and the apparent relevance of information to them depend crucially on perspective, and our approach to the regulation of transport is emphatically a function of what we take 'transport' to denote.

To illustrate this point, we might give further consideration to the possible role of RNA helicases in transport (chapter 5). Although there is no evidence that helicases are directly involved in actin binding within the nucleus, at least one and perhaps two or more such enzymes are required for messenger release from spliceosomes,[1,2] and another quite distinct helicase, P68, is present in nuclei and has some homology with eIF4A.[3,4] We recall that this initiation factor may be an actin-binding protein and is certainly capable of unwinding mRNA in an ATP-dependent manner.[5] Yet another, different RNA helicase is associated with the nuclear envelope and may be related to mRNA translocation. Because RNA-protein binding is likely to be influenced in terms of both specificity and affinity by RNA secondary structure, a successive arrangement of different helicases through nucleus and cytoplasm could be pertinent to a mechanism of transport of the kind outlined at the end of chapter 5. In Fig. 6.1 we illustrate the idea schematically; RNA molecules bound to either immobile (structural) or 'soluble' proteins form temporary linkages, constituting portions of the ephemeral 'solid-state' structure through which they migrate, and RNA helicases catalyse the formation and breakage of some of the links. This scheme is obviously speculative, but it is compatible with such evidence as exists and the idea seems never to have been entertained. It affords obvious possibilities for multiple controls and fine-tuning of RNA transport and on that basis (and especially given widespread interest in the biological roles of RNA helicases) it should merit consideration. The difficulty is that it is coherent only if the appropriate view of 'transport' is accepted. While 'transport' is seen as consisting only of translocation through the pore-complex, only the envelope-associated helicase can be perceived as potentially relevant.[6] The possibility that other helicases represent sites of transport control is concomitantly overlooked.

We claim no credibility for the scheme in Fig. 6.1 because there is no pertinent evidence. Our point is that until the perspective that allows such schemes becomes more widely adopted, the attention that is given to the regulation of nucleocytoplasmic RNA transport will be limited, potentially relevant information will be considered irrelevant, and such data as are obtained will be subjected to limiting and perhaps misleading interpretations. Similarly, an inappropriate perspective could distort our perception of the way in which protein transport is regulated.

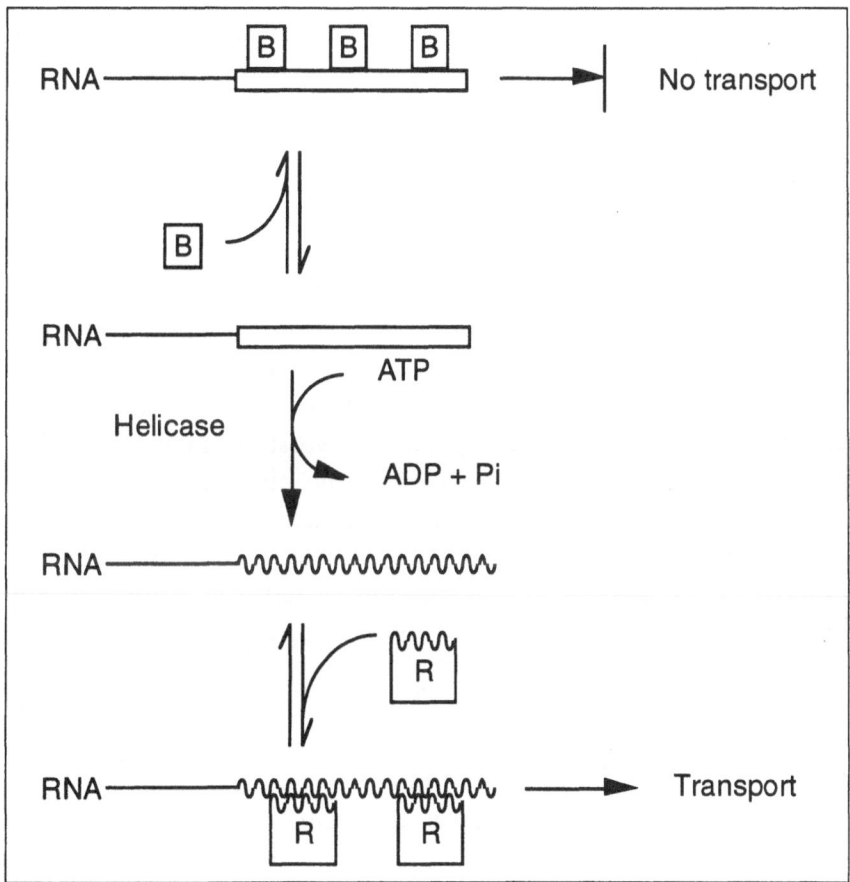

Fig. 6.1. Possible role of RNA helicases in transport. Secondary structure motifs in a particular region of an RNA model might determine protein affinities. By altering the local secondary structure, a helicase could decrease the affinity of the transportant for a blocking protein or transport inhibitor (B) and increase affinity for a transport receptor (R), binding being either direct or indirect. (Of course, the helicase could equally have the converse effect.)

CONTROL OF PROTEIN IMPORT: PHOSPHORYLATION AND CYTOPLASMIC ANCHORING

Certain proteins are cytoplasmically located for large portions of the cell cycle and enter the nucleus only at specific times. A consensus view has developed that the changed location is generally brought about by phosphorylation or dephosphorylation events. This is undoubtedly a common mechanism,[7] but it can operate in many ways. At least some proteins are cytoplasmically located when they are anchored to constitutively cytoplasmic components, and phosphorylation/dephosphorylation might provide one means of raising or dropping the anchors. Alternatively, changes in phosphorylation status might change the efficacy of a NLS. Whether anchoring or NLS efficiency is emphasized in a discussion of the data is likely to be influenced by the perspective adopted on nucleocytoplasmic transport.

For example, the Rel-related transcription factor NFκB is normally held in the cytoplasm as an inactive complex by IκB,[8] different isoforms of which might be rendered effective[9] or ineffective[10] by phosphorylation. The transcription factor NF-AT might be anchored in the cytoplasm of unstimulated T cells by an unidentified phosphorylated protein,[11] and although cAMP stimulates the nuclear import of NFIL this factor does not itself appear to be a substrate for protein kinase A.[12] On the other hand, interferon α activation of cells causes an apparently tyrosine kinase dependent uptake of the factor ISGF3a and here the phosphorylation target is the factor itself.[13-15] In *S. cerevisiae* the transcription factor SW15 is a substrate for the serine protein kinase cdc28, the yeast homolog of cdc2, and when it is phosphorylated it is cytoplasmically located.[16] Only in G1 is it sufficiently dephosphorylated to enter the nucleus, presumably because of mitotic cdc28 inactivation and concomitant dephosphorylation. Removal of the SW15 from the nucleus after G1 might involve intranuclear phosphorylation; in some vertebrate cells at least, protein kinases can gain access to the nuclear internum in response to certain stimuli.[17,18] The serine residues in SW15 that are phosphorylated by cdc28 are close to the nuclear location signal,[16] and it is possible that the imposition of a negative charge near the NLS inhibits LSB association. A similar mechanism seems to obtain in controlling the import of the tumorigenic factor v-jun in chick embryo fibroblasts,[19] and here again the phosphorylation is cell-cycle related. However, nuclear uptake of SV_{40} large-T antigen is markedly stimulated by casein kinase II dependent phosphorylation of a serine residue close to the NLS,[20] suggesting that the effect is not a simple electrostatic one. Uptake of SV_{40} large-T antigen seems to be dependent upon hsp70, which is not true of all karyophilic proteins (even ones with similar NLSs),[21] and it is possible that phosphorylation affects hsp70 interaction with the transportant-LSB complex in this case.

These examples show that phosphorylation of either transportant or cytoplasmic anchor can promote or inhibit nuclear import, depending on the cell and the karyophile. No simple or consistent picture emerges from the evidence. Four other points should be added that increase the complexity still further. First, phosphorylation of certain proteins (including at least two of M_r 60-62,000) present in digitonin-permeabilized cells seems to be a prerequisite for nuclear uptake of virtually any protein transportant in mammalian fibroblasts.[22] Second, when cell proliferation is accelerated by exogenous growth factors, several proteins are imported by the nucleus more or less simultaneously, suggesting some as yet elusive commonality of regulatory mechanism.[23] Third, the pore-complex itself changes during the cell cycle and there is as yet no persuasive evidence that these changes are phosphorylation-related. In BALB/c 3T3 cells both the translocation rate and the pore channel capacity were shown to peak around 1 hour after anaphase, and the channel size but not the translocation rate reached a minimum after 21 h.[24] Fourth, the nuclear import of glucocorticoid receptor requires dissociation from cytoplasmic adherents such as hsp 90, but here the requisite stimulus is not phosphorylation but addition of the hormone.[21]

The emphasis on phosphorylation/dephosphorylation is understandable in view of what we know about intracellular control and signal transduction mechanisms in general, but the absence of a coherent picture of nuclear protein import regulation indicates our current lack of understanding of this specific topic. In view of the intricacies implicit in the control of Ran functions by RCC1, Ran GAP and the Ran binding proteins (see chapter 4), it is a little surprising that the Ran/RCC1 system has not yet received major consideration as a target for control of nucleocytoplasmic protein and RNA fluxes. This is a development we might expect to see in the literature in the not too distant future.

Given that responses to extracellular stimuli and more particularly passage through the cell cycle require the up or down regulation of large numbers of genes, and that nuclear import of specific transcription factors and export of messengers play significant parts in such regulation, it would be surprising if the emerging picture of the control of nucleocytoplasmic transport were anything less than complicated. At present, this aspect of the field is in its infancy, and premature generalizations need to be resisted.[25]

EXTRACELLULAR PROTEINS CAN ENTER THE NUCLEUS

Some extracellular proteins can penetrate the cytoplasm, whether via endolysosomes or the endoplasmic reticulum or directly through the plasma membrane itself is unclear. Presumably proteins selected for such entry bind to cell surface receptors and are at least partially unfolded for transit through the bilayer. The general biological value of this mechanism is uncertain, but cell-penetrating proteins include some that are karyophilic; interleukin 1, fibroblast growth factor and the HIV-1 protein Tat are examples.[26-28] They also include glycoproteins,[29,30] which could have non-NLS dependent modes of uptake into the nucleus (chapter 4).

Why should this route be taken for information transfer rather than, say, a conventional cell surface binding followed by G-protein or tyrosine kinase activation? Certainly the internalized karyophiles seem to function as transcription factors,[26-30] but there seems no a priori reason why secreted signaling molecules should not activate endogenous transcription factors in target cells by mechanisms of the kind outlined in the previous section instead. Indeed, a mechanism for allowing external proteins into the cytoplasm can be subverted, as the effects of bacterial and plant protein toxins presumably demonstrate. It is unlikely to be more rapid or more efficient in terms of numbers of signaling molecules than transduction at the plasma membrane followed by kinase/phosphorylase modulation. Perhaps an internalized signal protein can adopt two states in the cytoplasm, one appropriate for nuclear import and the other for cytoplasmic storage and resecretion; this would allow the stimulated target cell to exert autocrine and paracrine effects after the original stimulus has been dissipated from the local environment, though why this should be advantageous in any particular case is not clear. At present we can only note that the phenomenon occurs, and observe that it represents another type of case in which transcriptional control does not appear to depend on the phosphorylation status of the relevant factor.

SIGNALING FROM THE EXTRACELLULAR MATRIX TO THE NUCLEUS

The notion that a solid-state information transduction system might link the cell's external environment to specific gene promoters is controversial, but an increasingly strong case can be made for it. Such a system could circumvent or supplement conventional biochemical signaling systems and would constitute a means of transporting information rather than particular molecules between the exterior, the cytoplasm and the nucleus.

The argument can be constructed from the following steps. First, the extracellular matrix (ECM) is known to alter the cytoskeleton by means of its connections through integrins, and second, alterations of the cytoskeleton can alter patterns of gene expression.[31] Third, the ECM is a major determinant of cell differentiation,[32] and specific gene expression is regulated by cell-ECM contact; albumin expression in hepatocytes is a well-established example[33-35] and it is far from unique. Fourth, the nuclear matrix undergoes dramatic compositional changes during differentiation,[36] including association with gene promoters[37] and transcription factors,[38] making it at least feasible that the in situ nucleoskeleton is similarly altered. Fifth, it is clear that new gene transcription entails chromatin reorganization at the nucleosome level[39] and during ECM-induced gene expression histone deacetylases become associated with nuclear matrix preparations.[40] In osteoblasts the nuclear matrix and ECM changes coincide temporally,[36-38] and in other cell types the nuclear matrix changes depend on the type of ECM with which the cells interact.[41] On the other hand, the transcription factors themselves are upregulated by ECM contact and this event, which is surely primary in differentiation, may well be the consequence of conventional biochemical signaling pathways.[34,35]

The concept of a continuous solid-state tissue matrix comprising ECM, cytoskeleton and nucleoskeleton is far from novel[42-45] and it has been argued that mechanical changes transmitted through this system could alter patterns of gene expression,[45] but the existence of functional in situ connections between cytoskeleton and nucleoskeleton has not been established unequivocally (see chapters 3 and

5).[46] The recent observation that nucleolin occurs in the cytoplasm and at the cell surface in mature and differentiating neurons and that it binds the differentiation-promoting IKVAV motif of the laminin α chain[47] lends further credence to the concept, however. It also supports the suggestion that nucleolin is identical with P110, given the apparently transcellular distribution of the latter (see chapter 5). On the other hand it is not clear how a purely mechanical system of connections can direct information from the cell surface to the expression of *specific* genes. This problem, the specificity problem, becomes dominant when we turn from the issue of 'information import' by the nucleus to that of the control of mRNA export.

THE SPECIFICITY PROBLEM

One of the difficulties in making sense of mRNA transport regulation lies in understanding how it can be specific enough to be biologically useful. It may well be the case that 5' methyl caps and 3' poly(A) tails are amongst the intramolecular signals necessary for transport (perhaps particularly for pore-complex transit) but they are very widespread amongst messengers—in the former case, universal—and as such afford no possibilities for species or family specific regulation of transport. Similarly, it is hard to see how the common NLSs of karyophilic proteins contribute to specific regulation of protein import. The AU_3A sequence that is present in one or many copies in mRNAs for certain cytokines (TNFα, IL1) and oncogene products (*fos, myc, sis*) and some transcription factors[48-50] probably binds particular proteins, at least two of which have been identified in some cell types. There is an AU-specific endoribonuclease that degrades messengers containing this motif[51,52] and a binding protein that blocks this enzyme and prolongs the half-lives of the mRNAs.[53-55] The latter seems selectively to accelerate export of AU_3A-containing messengers from the nucleus.[55] This might be evidence for selective transport of an mRNA family or class based on a specific intramolecular signal and an associated binding protein; the binding protein is probably inducible by appropriate extracellular stimuli, at least in macrophages.[52] But this is a single example, and family-selection is still far from individual molecular species selection.

It may be that transport is not in itself a selective process, though the arguments that have been offered to this effect presume a very restricted understanding of 'transport'.[56,57] On the other hand, it may be that individual selection is achieved through a series of family-specific signals of the AU_3A type, all shared on one molecule and each operating at a different step in transport (Fig. 6.2). It need not be assumed that AU_3A and its binding protein interact only (or at all) with the pore-complex machinery; enhanced nuclear export could be achieved by more efficient spliceosome detachment and/or migration through the nucleoskeleton. It should also be emphasized that signals might consist in

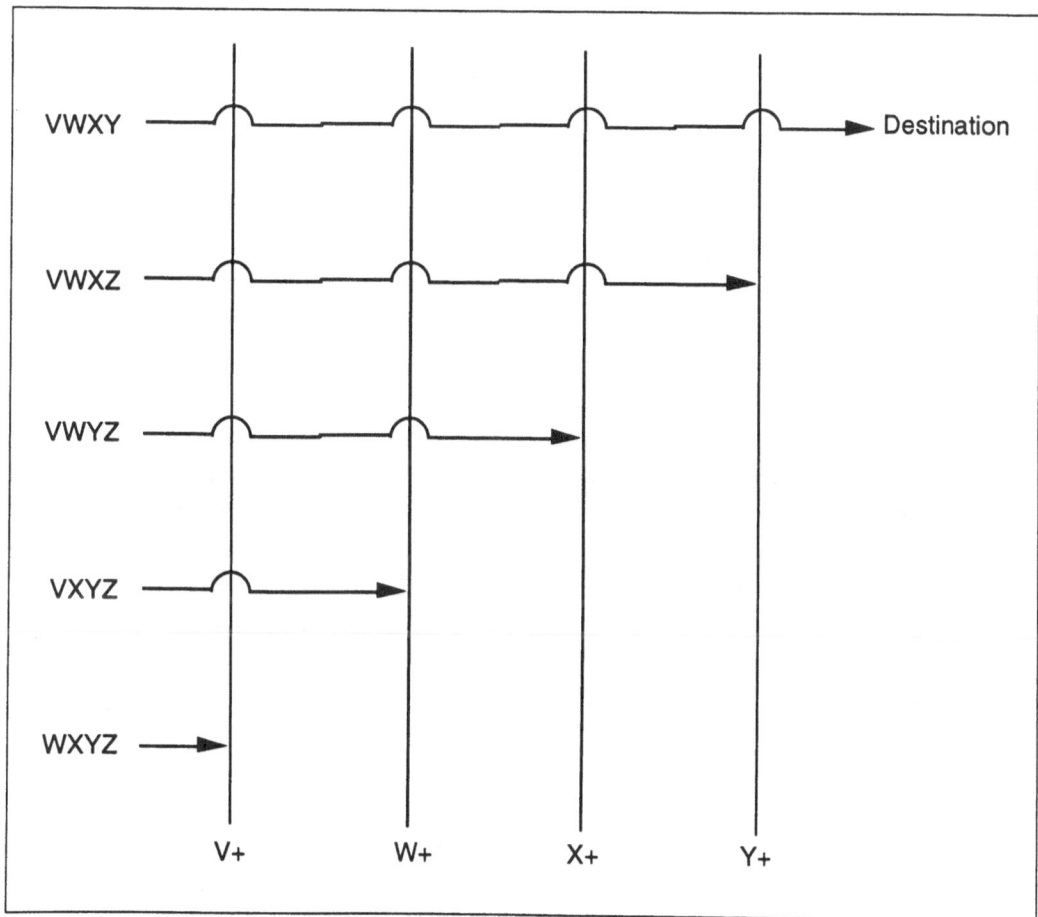

Fig. 6.2. An information filter model for RNA transport specificity. Each RNA molecule contains a four-element subset of five intramolecular signals V-Z. At successive stages in transport (e.g., nuclear packaging, nucleoskeletal, pore-complex, cytoskeletal) one of these signals is recognized. Although recognition of an individual widely-distributed signal does not in itself confer specificity on transport, a succession of such coarse information filters can in principle select out an individual molecular species for targeting the transport destination.

secondary structure conformations rather than primary sequence motifs, and here the potential modulatory role of RNA helicases might be recalled (see the introduction to this chapter).

INSULIN AND EPIDERMAL GROWTH FACTOR

The issues of specificity, location and mechanism appear again when we consider the sensitivity of RNA transport to extracellular stimuli. Insulin affects protein turnover by a variety of mechanisms depending on cell type, apparently modulating protein catabolism in some cases, translation in others, and mRNA transport in still others.[58] Presumably the insulin is internalized into the cytoplasm in the same way as fibroblast growth factor (see above). Efflux of mRNA from nuclei isolated from responsive cells exhibits a biphasic response with an optimum at around 10^{-11} M insulin, which is within physiological range, and there is a parallel stimulation of the nuclear envelope nucleoside triphophatase.[58,59] The activation of the enzyme seems to be the consequence of dephosphorylation of the poly(A) receptor P110.[59,60] This has the effect of decreasing the receptor affinity for poly(A)+ mRNAs, increasing the relative amount of high-abundance messengers transported to the cytoplasm and therefore promoting the output of major luxury proteins such as serum albumin.[61] So far as it goes, this account of insulin action on liver protein turnover accords with the evidence and is consistent with what we know about the RNA translocation machinery (chapter 5). However, it leaves us with a number of questions. First, hormone (including insulin) actions on target cells are specific, and it is not clear that the promotion of messenger transport merely on the basis of messenger abundance merits this description. In line with our argument in the previous section, this difficulty could be evaded if insulin also modulates other parts of the transport machinery, but this raises a second question: which other parts? In differentiated 3T3-442A adipocytes, insulin enhances phosphorylation of nucleolin by a mechanism that appears to be dependent on casein kinase II,[62] and again the response is biphasic and peaks at around 10^{-11} M hormone. Given the role of the nucleolus in pre-mRNA packaging and the early stages of transport this effect might be significant, but details remain obscure; however, the possible identity between P110 and nucleolin makes the mechanism credible.[63] As we noted in chapters 4 and 5, both are large RNA-binding phosphoproteins implicated in transport, both are highly labile to endogenous proteinases and might be self-cleaving,[64,65] and homologies between them might be significant for our understanding of transport overall. Their similar intracellular distributions were noted earlier in the present chapter.

Insulin is not alone in being reputedly able to modulate mRNA transport. Epidermal growth factor has the opposite effect; it promotes P110 phosphorylation,[59] possibly via protein kinase C α or β,[63] causing relative nucleoside triphosphatase inhibition and a tendency to increase the proportion of low-abundance messengers transported. The evidence so far as it goes seems unimpeachable, but the same problems remain: how is specificity achieved, and is there a multiplicity of targets within the overall transcellular transport machinery?

MRNA TRANSPORT IN ADENOVIRUS
AND INFLUENZA VIRUS INFECTED CELLS

Amongst the pathological changes that might throw light on the mechanism of mRNA transport and its regulation, the consequences of viral infection have received most attention during the past few years. To date there has been little systematic effort to relate these highly interesting findings to the remainder of the nucleocytoplasmic transport literature, and once again it is clear that a limited view of 'transport' has vitiated attempts at interpretation. Nuclear entry by virus components is a significant part of the viral infection process and is probably worth elucidating for that reason alone. Some viral proteins can enter the nucleus by virtue of their NLSs, which might be exposed and activated by limited proteolysis (e.g., in proteasomes) within the host cytoplasm; nucleic acids and other viral proteins might then be imported by 'hitchhiking' (chapter 4). However, many viral DNAs and associated proteins might be able to enter nuclei only during open mitosis, restricting their infectivities to actively dividing cells. Elucidation of viral nuclear transport mechanisms and their limitations might help to explain these differences.

The influenza virus protein NS1 has a high-affinity poly(A) binding domain, not wholly poly(A) specific. It is not absolutely required for exporting viral transcripts to the cytoplasm though it probably facilitates this process; in cells transfected with influenza virus genes but lacking NS1 the export of viral transcripts is about 25% as efficient as that in cells expressing the entire viral genome.[66]

NS1 inhibits the nucleocytoplasmic transport of host messengers but it has no apparent effect on splicing.[67] One possible interpretation of these results is that NS1 competes with P110 for the binding of host (polyadenylated) mRNAs and thereby sequesters them within the nucleus, making them unavailable for translocation;[68] in the absence of this protein, host and viral messengers compete at the pore-complex and viral mRNA transport becomes less efficient.[61] However, it is equally possible that NS1 subverts the intranuclear migration machinery in favor of viral transcripts, and when it is not present the viral mRNA is compelled to compete with host material for pathways of movement to the nuclear periphery. There is evidence that NS1 contains both an RNA binding domain and a protein binding domain.[69]

In late adenovirus infected cells, there is a more 'absolute' mechanism for ensuring that only viral and no host messenger reaches the cytoplasm.[70,71] Crucial for this switch is the complex of two viral proteins, E1B-E4,[72] located mainly at the viral replication-transcription sites in the nucleus. If either component is defective or if the complex is absent for any other reason then adenovirus mRNA is not moved efficiently from these sites towards the nuclear periphery,[73,74] and also host mRNA continues to be exported. Plainly the effect here is not primarily on translocation but on an earlier, intranuclear stage of mRNA transport ('release').[60] One suggestion is that E1B-E4 sequesters a soluble factor required for host messenger export.[67] However, when adenovirus-infected cells are superinfected with influenza virus, the transcripts of the latter reach the cytoplasm,[75] suggesting either that they do not require the soluble factor, or are able to recruit the factor despite the E1B-E4, or that the factor is not the correct explanation for the adenovirus block. E1B-E4 inhibits export in yeasts and concomitantly disrupts nuclear structure,[76] and adenovirus disperses nucleolar contents within host nuclei.[77] This suggests that the primary target of the E1B-E4 complex is the nucleoskeleton.

It is possible to accommodate both the influenza and adenovirus data within a scheme of the kind outlined in Fig. 6.1. Given transient solid-state components that bind poly(A) and are therefore susceptible to competition by NS1, and other components that bind RNAs via a 'soluble' coupling factor that is sequestered by E1B-E4, the effects of both type of infection on host mRNA transport can be explained in terms of intranuclear migration. If this view is broadly correct then the likeliest interpretation of the transfection data is that influenza virus messengers do not require the 'soluble' factor to migrate through the nucleoskeleton. This model directs our attention to the early stages of transport, not events at the pore-complex, as loci of viral disruption; an interpretation consistent with the opinion that the selection of messengers for export to the cytoplasm takes place at the release rather than the translocation stage.[57]

MESSENGER TRANSPORT IN HIV-INFECTED CELLS

As we have seen, NS1 affects mRNA transport apparently because it has specific RNA binding properties and possibly also because it leads to the formation of a 'viral nucleoskeleton' (allowing influenza virus messengers to bridge skeletal components but preventing host mRNAs from doing so). The three proteins produced early in HIV expression, Tat, Nef and Rev, seem to have analogous functions, particularly Rev, which binds to a specific viral mRNA sequence, the Rev-responsive element (RRE), the secondary structure of which it appears to stabilize.[78-83] The stabilization of a virus-specific nucleoskeleton complementary to the host one might also involve the Tat protein.[84,85] Tat, Nef and Rev are all trans-acting regulatory proteins encoded on the initial fully-spliced 2 kb transcript; significantly, Rev, a phosphoprotein of M_r 19,000, is concentrated in the nucleolus.[86-88] When they are present, the complete unspliced 9 kb and spliced 4 kb viral transcripts are made, encoding the remaining viral proteins.[89,90]

Rev is known to promote the export of RRE-containing RNAs to the cytoplasm in HIV-infected cells,[78,87,91] but the mechanism by which it does so has been debated. First, it probably increases the stability of such RNAs,[87] which could have marked consequences for cytoplasmic levels.[56] Second, its capacity to enable the export of unspliced RNAs might be related to inhibition of spliceosome assembly or promotion of disassembly.[92,93] This cannot be its only mode of action because Rev can enhance the export of RNAs lacking functional splice sites,[89,94] but changes in spliceosome stability could be related to migration through the viral nucleoskeleton[84,85] if this ephemeral structure is analogous to its host counterpart in being functionally coupled to splicing[95] (see chapter 5). The fact that the RRE is not absolutely required for its action may suggest that it can act directly on some part of the transport machinery without first binding to the RNA,[96] and indeed it does bind to the nuclear envelope-located P110 in vitro, decreasing the extent of poly(A)$^+$ mRNA binding and strongly inhibiting the nucleoside triphosphatase.[97] However, given: (a) the wide distribution of P110; (b) the concentration of Rev in the nucleolus and (c) the evidence for intranuclear action, this finding cannot be taken as indicating that 'Rev acts on mRNA transport' is tantamount to declaring 'Rev acts on the pore-complex'. Explicitly or implicitly, many contributors to research in this field have drawn exactly that inference.[78,94,97-99]

Like influenza virus and adenovirus transcripts, HIV-1 mRNAs inhibit the splicing and export of host premessengers and messengers, and Rev appears to modulate this action.[67] A possible explanation is that the virus-specific nucleoskeleton, in the formation of which Rev plays a part, acts as a medium for the migration as well as the splicing of viral transcripts, but its formation requires the recruitment of proteins that are also components of the host nucleoskeleton and the host mRNA transport machinery is thereby subverted. This account is compatible with the various apparently conflicting views of Rev action discussed above. The discovery by Pfeifer et al[97] that Rev binds to P110 could pertain to destabilization of the host nucleoskeleton as well as to inhibition of the translocation stage of transport. The Rev effector domain contains a nuclear export signal[100] and the nucleoporin-like receptor for it[101] might be identical with P110 or might be a P110 binding protein. Through this system Rev appears to be able to direct the export of any RRE-containing RNA.[102]

As in the cases of influenza virus and adenovirus, the effects of HIV infection on viral and host mRNA metabolism and transport seem more readily accommodated by the transport model outlined in chapter 5 than by any alternative. If this model is rejected, then the existing data seem to compel the inference that Rev, at least, has two or three quite different but simultaneous modes of action in HIV infected cells.

EFFECTS OF CARCINOGENS
ON MESSENGER TRANSPORT

Our interpretation of the viral effects on mRNA transport, like our view of the effects of insulin and epidermal growth factor, implies a coordinated response throughout the mRNA transport system. Changes in the nucleolus and nucleoskeleton are paralleled by changes in the translocation machinery, as indicated by alterations in nucleoside triphosphatase activity and the behavior of P110. Cytoplasmic aspects of transport might also be modified concomitantly though there seems to be no direct evidence for this. The simplest explanation for a set of coordinated responses is a commonalty of components and here again it is pertinent that neither the nucleoside triphosphatase[103] nor P110[104,105] is restricted to the nuclear envelope. A change in the level or activity of either or both of these molecules is likely to alter mRNA transport at the nucleoskeletal, nuclear envelope and possibly cytoskeletal levels simultaneously.

An early effect of carcinogen treatment is the increase in cytoplasmic mRNA complexity, associated with an increased rate of transport of messengers from nucleus to cytoplasm. This phenomenon has been known for more than two decades[106] and although it has not been shown to play a part in carcinogenesis it might herald a perturbation of cellular metabolism that is at least indirectly relevant to transformation. The increase of transport was shown by in vitro studies to be accompanied by activation of transport-promoting factors in the 100,000 g supernatant of cell homogenates,[107] and indeed different factors might appear in the cytoplasm during carcinogenesis.[108] These transport factors have been purified[108,109] and appear to be fragments of P110,[63] possibly generated by proteolysis or perhaps by translation of a family of messengers generated by differential splicing from a single gene transcript. Certainly carcinogen treatment seems to increase P110 levels.[110] As in the cases of virus infection we find a coordinated set of changes in nucleoside triphosphatase activity and levels of P110 and related proteins. It is likely that parallel changes occur in the nucleoskeleton, resulting in altered populations of RNAs destined for export; there are numerous hints in the 'nuclear matrix' literature that carcinogenesis is associated with structural changes within the nucleus. Much of the early evidence in this field was reviewed by Webb and his coworkers.[111]

What is far from clear is the connection between this constellation of carcinogen-induced alterations in mRNA transport and the actual induction of cancers through proto-oncogene activation. Presumably the key process is a transformation of the nucleoskeleton that adds proto-oncogenes to the set of transcribable DNA segments as well as 'scrambling' mRNA transport.

The chemical specificity of carcinogen effects on mRNA distribution can hardly be coincidental.[106] It might be supposed that long-range modifications of nuclear structure would have a wide variety of consequences, one subset of which could be changes in the pore-complex. Perhaps it is therefore not surprising that in some cancer cell lines the coupling of mRNA translocation to the nucleoside triphosphatase seems to be lost.[107,112]

A final observation for this section is that cocarcinogens are also potentially capable of altering the behavior of the mRNA transport system. For example, phorbol esters stimulate nuclear envelope-located protein kinase C, and by increasing the phosphorylation of P110[105] might directly increase the proportion of low abundance messengers in the cytoplasm.[61] In this case there might be no relevance to carcinogenesis; this effect of phorbol esters might be unrelated to the fact that they are cocarcinogens.

AGE-RELATED CHANGES IN MESSENGER TRANSPORT

The inference that any change in messenger RNA transport is multifaceted is particularly well-illustrated by the attenuation of the process that accompanies cell and tissue aging. There are age-related differences in levels of transcription, RNA processing, translation and RNA half-lives as well as in apparently all stages of transport,[113,114] and these differences pertain to total messenger populations and to specific messengers including those for liver albumin, cytochrome P450-LM2 and α_2-macroglobulin.[115-119]

Cells from old animals show 'relaxation' of gene expression. For example globin messengers appear in brain tissue, though they are not translated.[120] Also, immature premessengers are exported to the cytoplasm.[121] In view of our earlier discussions, these changes suggest that the nucleoskeletal organization of transcription, processing and intranuclear mRNA migration are altered with age. Certainly splicing appears to be impaired in cells from old animals, as illustrated by the case of ovalbumin gene transcripts in the hen oviduct,[122] and in the same tissue there is a concomitant fall of some 60% in the concentration of poly(A) binding proteins[123] that protect against endogenous ribonucleases.[124] Since the activity of at least one poly(A) metabolizing enzyme, the catabolic 2', 3'-exoribonuclease (EC 3.1-13.4), is considerably higher in old than in mature animals,[125] this suggests that poly(A) half-life diminishes with age. The data are consistent with this prediction.[126] Given that polyadenylation might be relevant both to splicing and to messenger stability,[127] this set of changes is likely to have implications for nuclear restriction, transport and turnover, with considerable repercussions for the population of translationally active polysomes in the cytoplasm.

There are concomitant changes in the translocation apparatus. In quail oviducts, nuclear envelope nucleoside triphosphatase activity increases during maturation or in response to estrogen treatment and decreases markedly again with aging or in response to estrogen withdrawal; the enzyme activity therefore parallels the level of transcription of the genes for the tissue's luxury proteins.[128] There are concomitant changes in the responsiveness of this enzyme in the nuclear envelope to poly(A) stimulation,[128,129] and in the activities of the endogenous kinase and phosphohydrolase that modulate the phosphorylation state of P110.[130] The phosphorylation state of P110 changes as expected, and there may be some decrease of P110 itself during aging.[131] The cytoplasmic/polysomal proteins (mRNA transport proteins) related to P110 (see previous section) also decrease markedly in older animals.[130,131]

It might be added that cytoskeletal changes in older animals are likely to relate to both poly(A)$^+$ mRNA anchoring and messenger half-life, and that these changes are probably coordinated with the rest of the mRNA metabolism and transport machinery.[131,132]

OVERVIEW

The brevity of this chapter has no doubt accentuated the heterogeneity of its contents but was necessitated by the generally undeveloped state of the topics reviewed. Only in respect of protein import control by phosphorylation and of viral effects on RNA export has significant progress been made during the past few years, and even in these cases our knowledge is not yet adequate for clear pictures to have emerged. Nevertheless, the weight of evidence, such as it is, suggests two general inferences. First, although modulation of transportant molecules can undoubtedly alter their ability to pass through the pore-complex, most control of nucleocytoplasmic transport is probably exerted at other stages and involves changes in anchoring within and migration through the cytoplasm or the nucleus. Second, while neither a strictly 'pore-complex only' nor a strictly 'solid state' perspective can comfortably accommodate all existing data on the regulation and pathologies of transport, nothing in the literature seems obviously incompatible with the compromise 'interactive' perspective developed in chapter 5. Therefore, we provisionally accept this perspective. It now remains to see what the prospects are for a model constructed within this perspective, and to what extent it can be generalized to other aspects of cell biology.

REFERENCES

1. Company M, Arenas J, Abelson J. Requirement of the RNA helicase-like protein PRP22 for release of messenger RNA from spliceosomes. Nature 1991; 349:487-488.
2. Schiver B, Guthrie C. PRP16 is an RNA-dependent ATPase that interacts transiently with the spliceosome. Nature 1991; 349: 494-496.
3. Iggo RD, Lane DP. Nuclear protein p68 is an RNA-dependent ATPase. EMBO J 1989; 8:1827-1836.
4. Hirling H, Scheffner M, Restle T et al. RNA helicase activity associated with the human p68 protein. Nature 1989; 339:562-563.
5. Ray BK, Lawson TG, Kramer JC et al. ATP-dependent unwinding of messenger RNA structure by eukaryotic initiation factors. J Biol Chem 1985; 260:7651-7653.
6. Schröder HC, Ugarkovic D, Langen P et al. Evidence for involvement of a nuclear envelope-associated RNA helicase activity in nucleocytoplasmic RNA transport. J Cell Physiol 1990; 145: 136-146.
7. Jans DA. The regulation of protein transport to the nucleus by phosphorylation. Biochem J 1995; 311:705-716.
8. Bauerle PA, Baltimore D. Activation of DNA binding activity in an apparently cytoplasmic precursor of the NF-κB transcription factor. Cell 1988; 53:211-217.
9. Link E, Kerr LD, Schreck R et al. Purified IκB-β is inactivated upon dephosphorylation. J Biol Chem 1992; 267:239-246.
10. Kerr LD, Inoue J, Davis N et al. The Rel-associated pp40 protein prevents DNA binding of Rel and NF-κB: relationship with IκB-β and regulation by phosphorylation. Genes Dev 1991; 5:1464-1476.
11 Liu J, Farmer JD, Lane WS et al. Calcineurin is a common target of cyclophilin-cyclosporin A and FKBP-FK506 complexes. Cell 1992; 66:807-815.
12. Metz R, Ziff E. cAMP stimulates the C/EBP-related transcription factor rNFIL-6 to trans-locate to the nucleus and induce *c-fos* transcription. Genes Dev 1991; 5:1754-1766.

13. Kessler DS, Levy DE. Protein kinase activity required for an early step in interferon-α signaling. J Biol Chem 1991; 266:23471-23476.

14. Fu X-Y. A transcription factor with SH2 and SH3 domains is directly activated by an interferon-α induced protein tyrosine kinase. Cell 1992; 70:323-335.

15. Schindler C, Shuai K, Prezioso VR et al. Interferon-dependent tyrosine phosphorylation of a latent cytoplasmic transcription factor. Science 1992; 257:809-813.

16. Moll T, Tebb G, Surana U et al. The role of phosphorylation and the CDC 28 protein kinase in the cell-cycle regulated nuclear import of the S cerevisiae transcription factor SWI5. Cell 1991; 66:743-758.

17. Nigg EA, Hilz H, Eppenburger HM et al. Rapid and reversible translocation of the catalytic subunit of cAMP-dependent protein kinase type II from the Golgi complex to the nucleus. EMBO J 1985; 4:2801-2806.

18. Chen RH, Sarnecki C, Blenis J. Nuclear localization and regulation of erk- and rsk-encoded protein kinases. Mol Cell Biol 1992; 12:915-927.

19. Tagawa T, Kuroki T, Vogt PK et al. The cell-cycle-dependent nuclear import of v-jun is regulated by phosphorylation of a serine adjacent to the nuclear localization signal. J Cell Biol 1995; 130:255-263.

20. Jans DA, Jans P. Negative charge at the casein kinase II site flanking the nuclear localization signal of the SV40 large T-antigen is mechanistically important for enhanced nuclear import. Oncogene 1994; 9:2961-2968.

21. Yang J, De Franco DB. Differential roles of heat shock protein 70 in the in vitro nuclear import of glucocorticoid receptor and simian virus 40 large tumor antigen. Mol Cell Biol 1994; 14:5088-5098.

22. Mishra K, Parnaik VK. Essential role of protein phosphorylation in nuclear transport. Exp Cell Res 1995; 216:124-134.

23. Bedells CH, Pennington SR. Analysis of growth factor stimulated nucleocytoplasmic protein transport using two-dimensional gel electrophoresis. Electrophoresis 1995; 16:1231-1239.

24. Feldherr CM, Akin D. Variations on signal-mediated nuclear transport during the cell cycle in BALB/c 3T3 cells. Exp Cell Res 1994; 215:206-210.

25. Powers M, Forbes DJ, Cytosolic factors in nuclear transport: What's importin? Cell 1994; 79:931-934.

26. Amalric F, Baldin V, BoscBierne I et al. Nuclear translocation of basic fibroblast growth factor. Ann NY Acad Sci 1991; 638:127-138.

27. Curtis BM, Widmer MB, Deroos P et al. IL-1 and its receptor are translocated to the nucleus. J Immunol 1990; 144:1295-1303.

28. Westendorp MO, Frank R, Ochsenbauer C. et al. Sensitization of

T cells to CD95-mediated apoptosis by HIV-1 Tat and gp120. Nature 1995; 375:497-500.

29. He J, Furmanski P. Sequence specificity and transcriptional activation in the binding of lactoferrin to DNA. Nature 1995; 373:721-724.

30. Modrell B, McDonald VL, Shoyab M. The interaction of amphiregulin with nuclei and putative nuclear localization sequence binding proteins. Growth Factors 1992; 7:305-314.

31. Blum JL, Wicha MS. Role of the cytoskeleton in laminin induced mammary gene expression. J Cell Physiol 1988; 135:13-22.

32. Adams JC, Watt FM. Regulation of development and differentiation by the extracellular matrix. Development 1993; 117:1183-1198.

33. Caron JM. Induction of albumin gene transcription in hepatocytes by extracellular matrix proteins. Mol Cell Biol 1990; 10:1239-1243.

34. Liu JK, DiPersio CM, Zaret KS. Extracellular signals that regulate liver transcription factors during hepatic differentiation in vitro. Mol Cell Biol 1991; 11:773-784.

35. DiPersio CM, Jackson DA, Zaret KS. The extracellular matrix coordinately modulates liver transcription factors and hepatocyte morphology. Mol Cell Biol 1991; 11:4405-4414.

36. Dworetzky SI, Wright KL, Fey EG et al. Sequence-specific DNA-binding proteins are components of a nuclear matrix-attachment site. Proc Natl Acad Sci USA 1992; 89:4178-4182.

37. Bidwell JP, Van Wijnen AJ, Fey EG et al. Osteocalcin gene promoter-binding factors are tissue-specific nuclear matrix components. Proc Natl Acad Sci USA 1993; 90:3162-3166.

38. Van Wijnen AJ, Bidwell JP, Fey EG et al. Nuclear matrix association of multiple sequence-specific DNA binding activities related to SP-1, ATF, CCAAT, C/EBP, OCT-1, and AP-1. Biochemistry 1993; 32:8397-8402.

39. McPherson CE, Shim EY, Freidman DS et al. An active tissue-specific enhancer and bound transcription factors existing in a precisely positioned nucleosomal array. Cell 1993; 75:387-398.

40. Hendzel MJ, Delcuve GP, Davie JR. Histone deacetylase is a component of the internal nuclear matrix. J Biol Chem 1991; 266:21936-21942.

41. Gordon JN, Shu WP, Schlussel RN et al. Altered extracellular matrices influence cellular processes and nuclear matrix organizations of overlying human bladder urothelial cells. Cancer Res 1993; 53:4971-4977.

42. Fey EG, Wan KM, Penman S. Epithelial cytoskeletal framework and nuclear matrix-intermediate filament scaffold: three-dimensional organization and protein composition. J Cell Biol 1984; 98: 1973-1984.

43. Bissett MJ, Hall HG, Parry G. How does the extracellular matrix direct gene expression? J Theoret Biol 1982; 99:31-68.

44. Pienta KJ, Coffey DS. Nuclear-cytoskeletal interactions: Evidence for physical connections between the nucleus and cell periphery and their alteration by transformation. J Cell Biochem 1992; 49:357-365.
45. Ingber DE. The riddle of morphogenesis: A question of solution chemistry or molecular cell engineering? Cell 1993; 75:1249-1252.
46. Georgatos SD. Towards an understanding of nuclear morphogenesis. J Cell Biochem 1994; 55:69-76.
47. Kibbey MC, Johnson B, Petryshyn R et al. A 110-kD nuclear shuttling protein, nucleolin, binds to the neurite-promoting IKVAV site of laminin-1. J Neurosci Res 1995; 42:314-322.
48. Caput D, Beutler B, Hartog K et al. Identification of a common nucleotide sequence in the 3' untranslated region of mRNA molecules specifying inflammatory mediators. Proc Natl Acad Sci USA 1986; 383:1670-1674.
49. Ryseck R-P, Hirai SI, Yanir M et al. Transcriptional activation of c-jun during G0/G1 transition in mouse fibroblasts. Nature 1988; 334:535-536.
50. Peppel K, Vinci JM, Baglioni C. The AU-rich sequences in the 3' untranslated region mediate the increased turnover of interferon mRNA introduced by glucocorticoids. J Exp Med 1991; 173:349-355.
51. Schröder HC, Dose K, Zahn RK et al. Isolation and characterization of the novel polyadenylate and polyuridylate-degrading endoribonuclease V from calf thymus. J Biol Chem 1980; 255:5108-5112.
52. Jochum C, Voth R, Rossol S et al. Immmunosuppressive function of hepatitis B antigens *in vitro*. Role of the endoribonuclease V as one potential *trans* inactivator for cytokines in macrophages and human hepatoma cells. J Virol 1990; 64:1956-1963.
53. Malter JS. Identifiction of an AUUUA-specific messenger RNA binding protein. Science 1989:246:664-665.
54. Gillis P, Malter JS. The adenosine uridine binding factor recognizes the AU-rich elements of cytokine, lymphokine and oncogene mRNAs. J Biol Chem 1991; 266:3172-3177.
55. Müller WEG, Slor H, Pfeifer K et al. Association of AUUUA-binding protein with A+U rich mRNA during nucleocytoplasmic transport. J Mol Biol 1992; 226:721-733.
56. Darnell JE. Variety in the level of gene control in eukaryotic cells. Nature 1982; 297:365-371.
57. Riedel N, Fasold H. Transport of ribosomal proteins and RNAs. In: Feldherr CM, ed. Nuclear Trafficking. San Diego: Academic Press, 1992:231-290.
58. Purrello F, Vigneri R, Clawson GA et al. Insulin stimulation of nucleoside triphosphatase activity in isolated nuclear envelopes. Science 1982; 216:1005-1007.

59. Schröder HC, Wenger R, Ugarkovic D et al. Differential effect of insulin and epidermal growth factor on mRNA translocation system and transport of specific poly(A)+mRNA and poly(A)-mRNA in isolated nuclei. Biochemistry 1990; 29:2368-2375.

60. Agutter PS. Nucleocytoplasmic transport of mRNA: its relationship to RNA metabolism, subcellular structures and other nucleocytoplasmic exchanges. In: Müller WEG, ed. Progress in Molecular and Subcellular Biochemistry. Vol 10. Heidelberg: Springer Verlag, 1988:16-96.

61. Schröder HC, Müller WEG, Agutter PS. Kinetic models for nucleocytoplasmic transport of messenger RNA. J Theoret Biol 1995; 174:169-177.

62. Csermely P, Schmaider T, Cheatham B et al. Insulin induces the phosphorylation of nucleolin—a possible mechanism for insulin-induced RNA efflux from nuclei. J Biol Chem 1993; 268: 9747-9752.

63. Schäfer P, Aitken SJM, Bachmann M et al. Immunological evidence for the localization of a 110 KDa poly(A) binding protein from rat liver in nuclear envelopes and its phosphorylation by protein kinase C. Cell Mol Biol 1993; 39:703-710.

64. Prochnow D, Thomson M, Schröder HC et al. Efflux of mRNA from resealed nuclear envelope ghosts. Arch Biochem Biophys 1994; 312:579-587.

65. Suzuki T, Suzuki N, Hosoya T. Limited proteolysis of rat liver nucleolin by endogenous proteases-effects of polyamines and histones. Biochem J 1993; 289:109-115.

66. Alonso-Caplan FV, Nemeroff ME, Qui Y et al. Nucleocytoplasmic transport: the influenza virus NS1 protein regulates the transport of spliced NS1 mRNA and its precursor NS1 mRNA. Genes Dev 1992; 6:255-267.

67. Krug RM, The regulation of export of mRNA from nucleus to cytoplasm. Curr Biol 1993; 5:944-949.

68. Qui Y, Krug RM. The influenza virus NS1 protein is a poly(A) binding protein that inhibits nuclear export of mRNAs containing poly(A). J Virol 1994; 68:2425-2432.

69. Qian XY, Alonso-Caplan FV, Krug RM. Two functional domains of the influenza virus NS1 protein are required for regulation of nuclear export of mRNA. J Virol 1994; 68:2433-2441.

70. Beltz GA, Flint SJ. Inhibition of HeLa cell protein synthesis during adenovirus infection: restriction of cellular mRNA sequences to the cytoplasm. J Mol Biol 1979; 131:353-373.

71. Pilder S, More M, Logan J et al. The adenovirus E1B-55K transforming polypeptide modulates transport or cytoplasmic stabilization of viral and host cell mRNAs. Mol Cell Biol 1986; 6:470-476.

72. Sarnow P, Hearing P, Anderson CW et al. Adenovirus early region 4 encodes functions required for efficient DNA replication, late gene expression and host cell shut off. J Virol 1984; 49:692-700.

73. Leppard KN, Shenk KT. The adenovirus E1B 55 kd protein influences mRNA transport via an intranuclear effect on RNA metabolism. EMBO J 1989; 8:2329-2336.

74. Ornelles DA, Shenk KT. Localization of the adenovirus early region 1B 55-kilodalton protein during lytic infection: association with nuclear viral inclusions requires the early region 4 34-kilodalton protein. J Virol 1991; 65:424-439.

75. Katze MG, Chen Y-T, Krug RM. Nucleocytoplasmic transport and VAI RNA-independent translation of influenza viral messenger RNAs in late adenovirus-infected cells. Cell 1984; 37:483-490.

76. Liang S, Hitami M, Tartakoff AM. Adenoviral E1B-55 KDa protein inhibits yeast mRNA export and perturbs nuclear structure. Proc Natl Acad Sci USA 1995; 92:7372-7375.

77. Puvion-Dutilleul F, Christiensen ME. Alterations of fibrillarin distribution and nucleolar ultrastructure induced by adenovirus infection. Eur J Cell Biol 1993; 61:168-176.

78. Malim MH, Hauber J, Le S-Y et al. The HIV-1 Rev *trans*-activator acts through a structured target sequence to activate nuclear export of unspliced viral mRNA. Nature 1989; 338:254-257.

79. Daly TJ, Cook KS, Gray GS et al. Specific binding of HIV-1 recombinant Rev protein to the Rev-responsive element *in vitro*. Nature 1989; 342:816-819.

80. Cochrane AW, Chen C-H, Rosen CA. Specific interaction of the human immunodeficiency virus Rev protein with a structured region in the *env* mRNA. Proc Natl Acad Sci USA 1990; 87: 1198-1202.

81. Daefler S, Klotman ME, Wong-Stahl F. *Trans*-activating Rev protein of the human immunodeficiency virus 1 interacts directly and specifically with its target RNA. Proc Natl Acad Sci USA 1990; 87:4571-4575.

82. Heaphy S, Dingwall C, Ernberg I et al. HIV regulator of virion expression (Rev) protein binds to an RNA stem-loop structure located within the Rev response element region. Cell 1990; 60:685-693.

83. Kjems J, Brown M, Chang DD et al. Structural analysis of the interaction between the human immunodeficiency virus Rev protein and the Rev response element. Proc Natl Acad Sci USA 1991; 88:683-687.

84. Heaphy S, Finch JT, Gait MJ et al. Human immunodeficiency virus type 1 regulator of virion expression, Rev, forms nucleoprotein filaments after binding to a protein-rich "bubble" located within the Rev-responsive region of viral mRNAs. Proc Natl Acad Sci USA 1991; 88:7366-7370.

85. Müller WEG, Okamoto T, Reuter P et al. Functional characterization of Tat protein from human immunodeficiency virus: evidence that Tat links viral RNAs to nuclear matrix. J Biol Chem 1990; 3803-3808.

86. Cullen BR, Hauber J, Campbell K et al. Subcellular location of the human immunodeficiency virus *trans*-acting *art* gene product. J Virol 1988; 62:2498-2501.

87. Felber BK, Hadzopoulou-Cladaras M, Cladaras C et al. Rev protein of human immunodeficiency virus type 1 affects the stability and transport of the viral mRNA. Proc Natl Acad Sci USA 1989; 86:1495-1499.

88. Cochrane AW, Perkins A, Rosen CA. Identification of sequences important in the nucleolar localization of human immunodeficiency virus Rev. Relevance of nucleolar localization to function. J Virol 1990; 64:881-885.

89. Rosen CA, Terwilliger E, Drayton A et al. Intragenic *cis*-acting *art* gene-responsive sequences of the human immunodeficiency virus. Proc Natl Acad Sci USA 1988; 85:2071-2075.

90. Kim S, Byrn R, Groopman J et al. Temporal aspects of DNA and RNA synthesis during human immunodeficiency virus infection: evidence for differential gene expression. J Virol 1989; 63: 3708-3713.

91. Hadzopoulou-Cladaras M, Felber BK, Cladaras C et al. Rev protein of human immunodeficiency virus type 1 affects viral RNA and protein expression via a *cis*-acting sequence in the *env* region. J Virol 1989; 63:1265-1274.

92. Cheng DD, Sharp PA. Regulation by HIV Rev protein depends upon recognition of splice sites. Cell 1989; 59:789-795.

93. Lu S, Heimer J, Rekosh D et al. U1 small nuclear RNA plays a direct role in the formation of a Rev-regulated human immunodeficiency virus *env* mRNA that remains unspliced. Proc Natl Acad Sci USA 1990; 87:7598-7602.

94. Emerman M, Vazeux R, Peden K. The Rev gene product of the human immunodeficiency virus affects envelope-specific RNA localization. Cell 1989; 57:1155-1165.

95. Zeitlin S, Parent A, Silverstein S et al. Pre-mRNA splicing and the nuclear matrix. Mol Cell Biol 1987; 7:111-120.

96. Venkatesan S, Gerstiberger SM, Park H et al. Human immunodeficiency virus type 1 Rev activation can be achieved without Rev-responsive element RNA if Rev is directed to the target as a Rev-MS2 fusion protein which tethers the MS2 operator RNA. J Virol 1992; 66:7469-7480.

97. Pfeifer K, Weiler BE, Ugarkovic D et al. Evidence for a direct interaction of Rev protein with nuclear envelope mRNA translocation system. Eur J Biochem 1991; 199:53-64.

98. Cullen BR, Malim MH. The HIV-1 Rev protein: prototype of a novel class of eukaryotic post-transcriptional regulators. Trends Biochem Sci 1991; 16:346-360.

99. Rosen CA. Regulation of HIV gene expression by RNA-protein interactions. Trends Genet 1991; 7:9-14.

100. Fischer V, Huber J, Boelens WC et al. The HIV-1 Rev activation domain is a nuclear export signal that accesses an export pathway used by specific cellular RNAs. Cell 1995; 82:475-483.

101. Fritz CC, Zapp ML, Green MR. A human nucleoporin-like protein that specifically interacts with HIV Rev. Nature 1995; 376:530-533.

102. Fischer V, Meyer S, Teufel M et al. Evidence that HIV-1 Rev directly promotes the nuclear export of unspliced RNA. EMBO J 1994; 13:4105-4112.

103. Clawson GA, Song Y-L, Schwartz AM et al. Interaction of human-immunodeficiency virus type 1 rev protein with nuclear scaffold nucleoside triphosphatase activity. Cell Growth Different 1991; 2:575-582.

104. Schweiger A, Kostka G. Concentration of particular high molecular mass phosphoprotein in rat liver nuclei and nuclear matrix decreases following inhibition of RNA synthesis with α-amanitin. Biochim Biophys Acta 1984; 782:262-268.

105. Schröder HC, Diehl-Seifert B, Rottmann M et al. Functional dissection of nuclear envelope mRNA translocation system: effects of phorbol esters and a monoclonal antibody recognizing cytoskeletal structures. Arch Biochem Biophys 1988; 261:394-404.

106. Shearer RW. Specificty of chemical modification of RNA transport by liver carcinogens in the rat. Biochemistry 1974; 13:1764-1769.

107. Schumm DE, Hananausek-Walasak M, Yannarell A et al. Changes in nuclear RNA transport incident to carcinogenesis. Eur J Cancer 1977; 13:139-147.

108. Moffett RB, Webb TE. Characterization of a messenger RNA transport protein. Biochim Biophys Acta 1983; 740:231-242.

109. Schröder HC, Rottmann M, Bachmann M et al. Proteins from rat liver cytosol which stimulate mRNA transport: purification and interactions with the mRNA transport system. Eur J Biochem 1986; 159:51-59.

110. Clawson GA, Koplitz M, Moody DE et al. Effects of thioacetamide treatment on nuclear envelope nucleoside triphosphatase activity and transport of RNA from rat liver nuclei. Cancer Res 1980; 40:75-79.

111. Palayoor T, Schumm DE, Webb TE. Transport of functional messenger RNA from liver nuclei in a reconstituted cell-free system. Biochim Biophys Acta 1981; 654:201-210.

112. Otegui C, Patterson RJ. RNA metabolism in isolated nuclei: processing and transport of immunoglobulin light chain sequences. Nucleic Acids Res 1981; 9:4767-4781.

113. Moore RE, Goldsworthy TL, Pitot HC. Turnover of 3'-polyadenylate containing RNA in livers from aged, partially hepatectomized, neonatal, and Morris 5123C hepatoma-bearing cells. Cancer Res 1980; 40:1449-1454.

114. Schröder HC, Müller WEG. Age-correlated decrease in the nuclear

restriction of mRNA. In: Robert L, Hofecher G, eds. The Theoretical Basis of Aging Research. Proceedings of the 7th Wiener Symposium on Experimental Gerontology. Wien: Facultas, 1990:
123-132.

115. Horbach GJMJ, Princen HMG, van der Kroef M et al. Changes
in the sequence content of albumin mRNA and in its translational
activity in the rat liver with age. Biochim Biophys Acta 1984;
783:60-66.

116. Dilella AG, Chiang JYL, Steggles AW. The quantitation of liver
cytochrome P450-LM2 mRNA in rabbits of different ages and after phenobarbital treatment. Mech Aging Dev 1982; 19:113-125.

117. Horbach GJMJ, van der Boom H, van Bezooijen CF et al. Molecular aspects of age-related changes in albumin synthesis in fe-
male WAG/Rij rats. Life Sci 1988; 43:1707-1714.

118. Murty CV, Mancini MA, Chatterjee B et al. Changes in transcriptional activity and matrix association of $\alpha_{(2u)}$-globulin gene family
in the rat liver during maturation and aging. Biochim Biophys Acta
1988; 949:27-34.

119. Rath PC, Kanungo MS. Age-related changes in the expression of
cytochrome P450(b+e) gene in the rate after phenobarbitone ad-
ministration. Biochem Biophys Res Commun 1988; 157:1403-1409.

120. Ono T, Cutler RG. Age-dependent relaxation of gene expression:
increase of endogenous murine leukaemia virus-related and globin-
related RNA in brain and liver of mice. Proc Natl Acad Sci USA
1978; 75:4431-4435.

121. Schröder HC, Trölltsch D, Friese U et al. Mature mRNA is selectively released from the nuclear matix by an ATP/dATP-dependent
mechanism sensitive to topoisomerase inhibitors. J Biol Chem 1987;
262:8917-8925.

122. Schröder HC, Messer R, Breter HJ et al. Evidence of ovalbumin
heterogeneous nuclear RNA processing in hen oviduct. Mech Aging Dev 1985; 30:319-324.

123. Bernd A, Batke E, Zahn RK et al. Age-dependent gene induction
in quail oviduct. XV.Alterations of the poly(A)-associated protein
pattern and of the poly(A) chain length of mRNA. Mech Aging
Dev 1982; 19:361-377.

124. Müller WEG, Arendes J, Zahn RK et al. Control of enzymic hydrolysis of poly(A) segment of messenger RNA: role of poly-
adenylate-associated proteins. Eur J Biochem 1978; 86:283-290.

125. Müller WEG, Zahn RK, Schröder HC et al. Age-dependent enzy-
matic poly(A) metabolism in quail oviduct. Gerontology 1979;
25:61-68.

126. Schröder HC, Schenk P, Baydoun H et al. Occurrence of short-
size oligo (A) fragments during course of cell cycle and aging. Arch
Gerontol Geriat 1983; 2:349-360.

127. Carlin RK. The poly(A) segment of mRNA: (1) Evolution and

function and (2) the evolution of viruses. J Theor Biol 1978; 71:323-334.

128. Bernd A, Schröder HC, Lehhausen G et al. Alteration of activity of nuclear envelope nuceloside triphosphatase in quail oviduct and liver in dependence on physiological factors. Gerontology 1983; 29:394-398.

129. Bernd A, Schröder HC, Zahn RK et al. Age dependence of polyadenylate-stimulation of nuclear envelope nucleoside triphosphatase. Mech Aging Dev 1982; 20:331-341.

130. Yannarell A, Schumm DE, Webb TE. Age dependence of nuclear RNA processing. Mech Aging Dev 1977; 6:259-264.

131. Schröder HC, Becker R, Bachmann M et al. Differential changes of nuclear envelope associated enzyme activities involved in nucleocytoplasmic mRNA transport in the developing rat brain and liver. Biochim Biophys Acta 1986; 868:108-118.

132. Schröder HC, Zahn RK, Müller WEG. Role of actin and tubulin in the regulation of poly(A) polymerase-endoribonuclease IV complex from calf thymus. J Biol Chem 1982; 257:2305-2309.

CHAPTER 7

EVALUATION OF A MODEL

INTRODUCTION

The thesis and antithesis that have formed the recurrent theme of this book, i.e., the pore-complex dominated and the solid-state perspectives on nucleocytoplasmic transport, have been explored in some detail. However, it would be glib to suggest that the interactive perspective discussed in chapter 5 and 6 represents a synthesis or resolution of these ideas, no matter how much circumstantial support accrues to it from existing experimental data. Certainly it envisages transportants as neither freely mobile nor bound to permanent fibrillar structures during transit, which seems like a compromise between the main perspectives, but this misses a crucial qualitative difference between them. The pore-complex only view of transport is traditionally 'biochemical' in the sense that it exclusively addresses defined structures with definite compositions, and therefore implies that fully detailed knowledge of the components will elucidate function as clearly as we might wish. In other words, sufficient progress in biochemical morphology will lead with little further ado to a complete understanding of nucleocytoplasmic transport. This is in line with the style of thinking that typified traditional biochemistry and remains explicit in contemporary molecular biology. In contrast, the solid-state perspective is traditionally 'biophysical'. Its concern is far more with the physical state of multimolecular assemblages than with the chemical details of individual components and their interactions. This is not to say that a solid-state transport system cannot be usefully elucidated by advances in biochemical morphology, because obviously it can. Rather, it is to say that biochemical morphology alone cannot provide a complete understanding of the transport process(es) because what matters crucially are irreducible statements about the physical state and properties of the large-scale structures involved. In this respect, the perspective is misaligned with the dominant thinking style in molecular biology; and, in this fundamental characteristic, the new perspective we have outlined in this book resembles the solid-state more than the pore-complex-only viewpoint (Fig. 7.1). In case an apology seems to be needed, we remind the reader that transport processes in general are matters of physics, not chemistry.

We begin this final chapter with a reconsideration of the attempts to express the solid-state perspective through more specific models.[1] Each of these attempts is problematic,[1] and it is now appropriate to ask whether the new perspective addresses the difficulties satisfactorily. After this we explore the applicabilities of this and alternative perspectives to intracellular transport in general, and consider the possible resolution of conflicts with dominant thinking styles in modern subcellular biology.

SOLID-STATE TRANSPORT: ARGUMENTS AND MODELS

The arguments in favor of a solid-state view of RNA transport came from many lines of evidence. The association of transcription with the nucleoskeleton[2] and of processing with this same structure[3] were suggestive, and the inference that intranuclear RNA is not freely mobile has been supported by both in situ[4-7] and in vitro[8-10] findings. Nevertheless, the interpretation of these findings has been challenged.[11,12] General and specific associations between mRNAs and the cytoskeleton have been discovered and reviewed.[13-15] Isolated nuclei have been shown to maintain normal RNA restriction unless they swell or proteolysis is allowed,[16,17] but again the interpretation is not straightforward (see, for example, ref. 12). The solid-state idea might also apply to protein transport, given the nondiffusive characters of many cytoplasmic and nuclear proteins[1,18-20] and the extreme nucleocytoplasmic concentration ratios of some,[21] and the notion is supported by the definite tracks followed by some proteins in the nucleus[20] and possibly the cytoplasm.[14] All this evidence has been discussed in earlier chapters in this book.

Three models that have been envisaged and peripherally discussed (or tacitly assumed) in the field were articulated and reviewed previously.[1] They were given the names 'direct transfer',

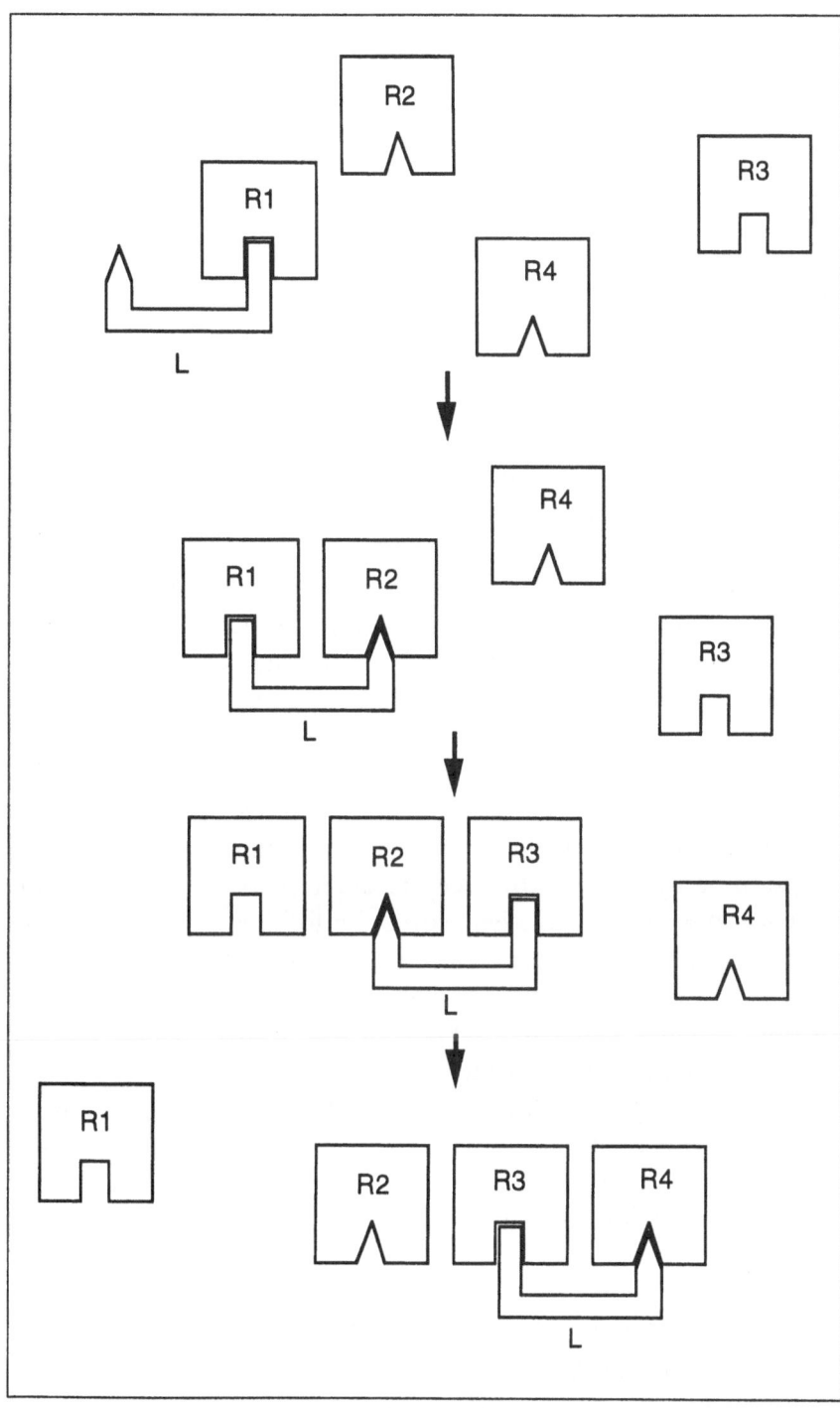

Fig. 7.1. A model corresponding to the 'interactive' perspective. A set of mobile receptors R1-R4 forms an immobile array transiently when ligand L crosslinks subsets of them. In this way a solid-state transport framework is induced by the ligand, but might not be constitutively present in the cell.

'motor driven' and 'assembly driven' (Fig. 7.2). The first is analogous with metabolic channeling,[22,23] the second with axonal transport and the third with cytoskeletal fibril dynamics.

The direct transfer model bears the closest resemblance to the perspective evolved in this book since, unlike the others, it does not presuppose the existence of a continuous fibrillar system extending between nucleus and cytoplasm. Instead, it envisages the transportant species being temporarily immobilized, much as successive intermediates are temporarily immobilized as they pass between the active sites of serially-associated glycolytic enzymes in the cytoplasmic compartment. The direct transport model is readily fitted to some bodies of data, such as the lack of explicit ATP requirement for some stages of transport, the capacity of some proteins and RNAs to shuttle between nucleus and cytoplasm and the web-like intranuclear distributions of some intranuclear messengers and their precursors.[1,12] It does not readily explain the vectorial character of some transport processes unless this is specially conferred by the pore-complex,[24,25] and it cannot account for defined intranuclear tracks[6,7] or specific associations between mRNAs and the cytoskeleton[15] without ad hoc modification in respect of individual cases. Also, it gives no clear predictions that can be critically tested, even in principle. The 'interactive' perspective outlined in chapter 5 suggests that the direct transfer model should be modified by making the receptor alignments stable only in the presence of a ligand; in other words, by linking different receptor molecules by the transportant itself. This loses none of the advantages of the model but vectorial migration can still be accounted for only through some such scheme as that shown in Fig. 5.3, and it remains unclear why definite tracks should be apparent in some cases, unless migration kinetics (particularly the rate constants of receptor-ligand dissociations) are particularly slow. This is at least a testable hypothesis in principle, and may be investigated in practice if it is provisionally assumed that the 'receptors' include NuMA and actin.

MOTOR DRIVEN AND ASSEMBLY DRIVEN MODELS: PRO AND CONTRA

The difference between these models lies in the nature of the association between transportant and fibrils. If transport is motor driven then attachment is via a motor that generates relative

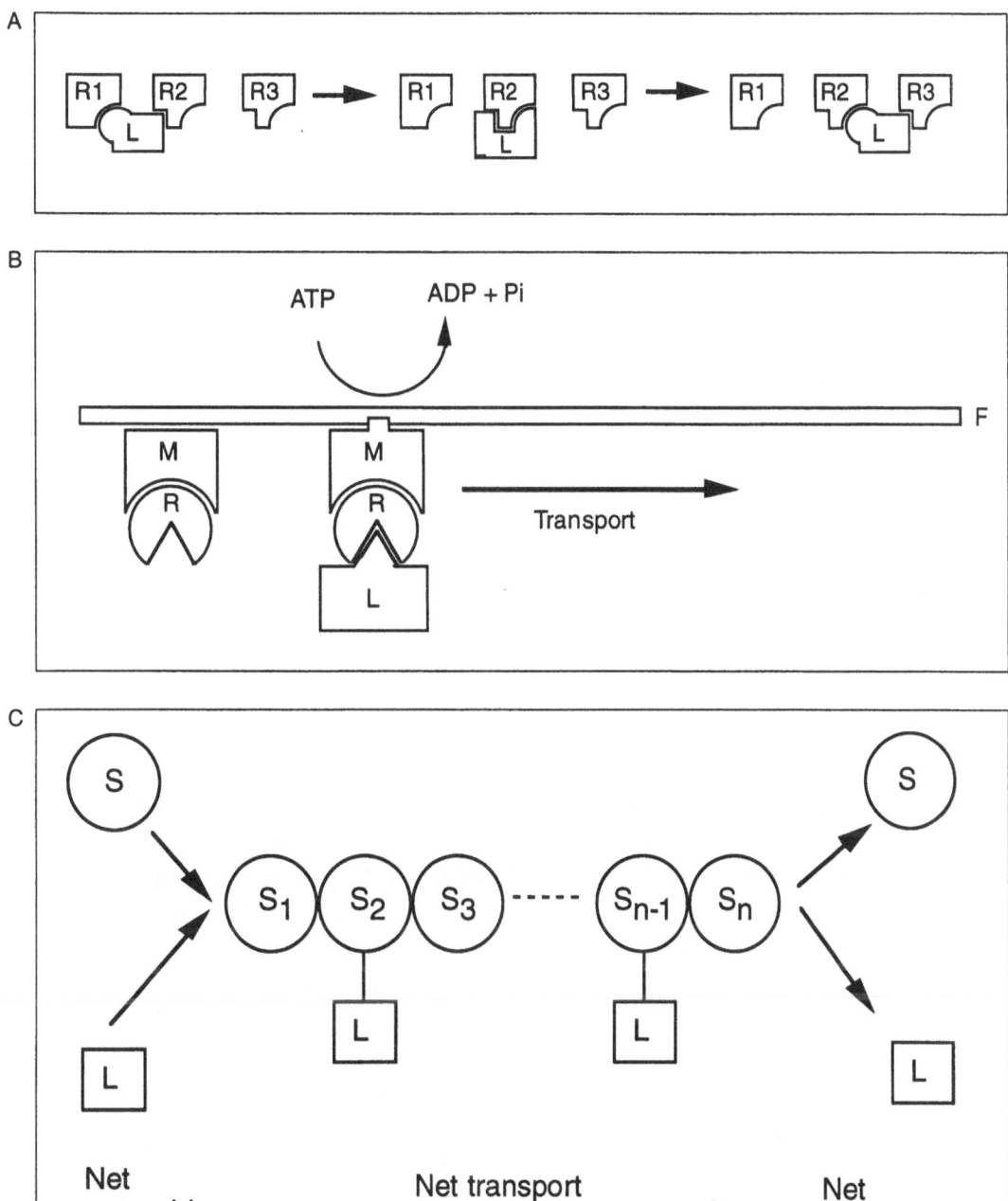

Fig. 7.2. Three models for solid-state transport. (A) Direct transfer. A sequence of receptors R1, R2... is arranged so that the ligand, L, is linked between each successive pair. During movement the ligand can be modified (as in RNA processing). (B) Motor driven. The receptor (R) is associated with a motor (M) which generates movement along a fibril (F) when the ligand is bound. (C) Assembly driven. The fibril is a linear array of subunits, S, some but not necessarily all of which bind the ligand (L) during assembly.

motion. If it is assembly driven then attachment to the fibril is fixed and movement is generated by fibril growth or retraction. Both models conform to experimental data in respect of certain general and specific features; ATP requirement, the involvements of actin in nucleocytoplasmic transport, the cytoskeletal attachments of (most) polysomes, the apparent ubiquity of P110 and its relatives in the cell, the vectorial character of most transport processes and the intranuclear tracks associated with some. The assembly driven model might also account for the 'free' or 'soluble' fraction of cytoplasmic polysomes, the size of which presumably depends inter alia on the rate of fibril disassembly; it might even suggest a correlation between the mRNA transport rate and fibril turnover. If fibrils can extend in any direction from a single nucleation point then nucleocytoplasmic shuttling and even the web-like intranuclear distributions of many messengers could be accounted for by an assembly-driven process.

On the other hand, both models, especially the motor driven model, presuppose a continuous if perhaps dynamic system of fibrils extending through nucleus and cytoplasm. As we have seen, this is at least possible, though the direct electron microscopic evidence[26] is not easy to interpret.[27] However, the fibrils must be different in different parts of the system. There seems to be no fibril-forming, functional isoform of NuMA in the interphase cytoplasm, and the components of the pore-complex cytoplasmic fibrils (chapter 3) are apparently absent from the rest of the proposed network. This makes the models at best cumbersome, since different but coordinated motors or assembly-disassembly mechanisms have to be proposed for different regions. It is also clear that the properties of the pore-complex nucleoside triphosphatase are different from those of any known motor.[14] Some transport stages such as the cytoplasmic migration of proteins to the nuclear surface and probably the outward cytoplasmic movement of messengers are apparently (mostly) ATP-independent, which is difficult to accommodate to a motor or assembly driven scheme. It is very difficult to reconcile a motor-driven model with the appearance of 'extra-chromosomal networks'[12] and both these models presuppose high-affinity receptor-ligand binding, in contrast to the direct transfer model, according to which no receptor can bind very tightly to the ligand. In this last respect the direct transfer model accords more satisfactorily with the data. Finally, the assembly driven model resists any formulation that fits it both to nucleocytoplasmic transport data and to what is known about assembly-disassembly mechanisms.

This balance of pro and contra arguments indicates that neither of the models is acceptable in the simple forms in which they were proposed.[1] It is nevertheless possible to assimilate them as partial models into the interactive perspective. For instance, if one receptor in a sequence of the kind illustrated in Fig. 7.1 chanced to be tightly associated with a fibrillar element of the cytoskeleton or with a motor operating on that element, then some portion of the overall transport process could be in effect motor or assembly driven. By this type of reasoning, it is formally possible to rescue the advantages of either motor or assembly driven models while evading their disadvantages. Whether the resulting hybrid scheme bears any significant resemblance to reality is, of course, quite another matter. Nevertheless, something of the kind must happen at least at the level of the pore-complex, where transportants must become indirectly bound to more or less permanent fibrils including the p45-p54-p58-p62 complexes of the central plug (chapter 4).

TESTING A 'CONSENSUS MODEL' OF NUCLEOCYTOPLASMIC TRANSPORT

The 'interactive perspective' immediately suggests a model comprising receptors R_1, R_2,...R_x, a ligand which might be modified during transport (as RNA is processed, for example), and fibrils with life expectancies several orders of magnitude greater than those of any individual receptor-ligand complex. This picture is a hybrid between those of Fig. 7.1 and 7.2(a). It suggests an approach to the biochemical morphology of the nucleocytoplasmic transport machinery overall, if 'biochemical morphology' is an apt phrase for a system with no durable structure; the overall strategy has to be based on the use of one receptor to identify others that are its nearest functional neighbors. One tactic might be to crosslink, say, messengers with bound proteins and thereby to isolate candidate receptors.[28] Alternatively, a column-immobilized receptor should bind its nearest functional neighbors in the presence but not the absence of ligand. Specific (blocking) antibodies against candidate receptors identified by such means must be shown to inhibit transport in situ, at least in permeabilized cells. Since some receptors are already known (e.g., importin 60 and its homologues for proteins, P110 and the CBPs for mRNAs, pore-complex p62 for both classes of transportant; see chapters 4 and 5), there is an extant basis for a research program of this kind.

An alternative strategy lies in the growing yeast genetic libraries and the techniques of synthetic lethal mutant screening (chapter 4). For instance, using a strain carrying a nonlethal mutation on a pore-complex protein known to be involved in mRNA transport, a sufficiently large screening range should identify other receptors outside the pore-complex. The main difficulty might be in identifying homologues in vertebrates and other 'higher' eukaryotes. This will remain a barrier to progress until the parallels between yeast and vertebrate pore-complexes in respect of biochemical morphology have been clarified.

Further details of possible future research programs would be egregious. The point we wish to make is that a model of the kind we have proposed is heuristically productive; it directly yields experimentally testable hypotheses and affords the prospect of underpinning the emergence of an increasingly detailed and increasingly credible understanding of nucleocytoplasmic transport processes. In this respect, the solid-state transport concept has always been scientifically defective, just as the pore-complex-only presumption has been scientifically naive (or, to be less euphemistic, wrong).

REFLECTIONS ON KINETICS AND TOPOLOGY

A general kinetic equation for a model of the type we have discussed could be deduced but it would serve little purpose at this stage. Instead, a qualitative consideration of the model's demands would be more useful, in so far as it would clarify the process by which a transportant is moved at a viable rate. We shall consider a generalized single step in transport, during which the transportant (ligand) ceases to be linked between receptors A and B and is linked instead between receptors B and C. (The receptors need not be biochemically distinct; in the simplest case they could all be of the same molecular type. But they do need at least temporarily to occupy different locations.) Obviously the probability of this event depends on the number of A-ligand-B complexes and the number of available C receptors. It also depends on mechanistic details, and here we can distinguish two possibilities.

First, suppose that C is itself mobile, i.e., is not firmly linked to a fibril or other structure. The chance of transfer from A-link to C-link now depends on the accessibility of the A-ligand junction to C. Overall, transfer probability will depend on a product of these three parameters:

p(transfer) $\propto$ (# of AB links)•(# of C)•(access of C to A-link)

However, the *rate* of transfer does not depend only on this probability. It also depends on the reorientation of the B-ligand-C complex. Nothing would be gained if C simply replaced A at the same location; C has to move, with the ligand attached, to a point distant from A in molecular terms and in the direction of overall transport. In view of the probable sizes of the species involved, it seems likely that this step will in most cases be very much slower than the individual ligand association and dissociation events and will therefore be the main determinant of the transfer rate for a given transfer probability.

Second, suppose that C is effectively immobile, being bound to some structure such as a fibril or a pore-complex. Access of C to the A-link of the ligand no longer enters our considerations. Instead, the formation of the B-ligand-C complex depends on detachment of the ligand from A, spatial reorientation of the ligand, and binding to C. The reorientation step precedes rather than succeeds the formation of the new complex in this case, so

p(transfer) $\propto$ (# of AB links)•(# of C)•(ligand reorientation).

Although there is no compelling reason to suppose that appropriate ligand reorientation will be very rapid, so that the expected rate of this event would be higher than the rate of C-access to the A-ligand site, we now at least eliminate the additional (slow) reorientation step after transfer. We can also assume on the basis of size that the reorientation of ligand alone will normally be faster than reorientation of ligand with C bound, as in the previous case.

This argument suggests that, as a rule, we should expect macromolecule transport between nucleus and cytoplasm to be faster, and more efficient in at least the kinetic sense, if the successive receptors are immobilized than if they are mobile. Our new model therefore becomes close in spirit to a solid-state mechanism, even though it does not presume the existence of a complete framework for transport (fibrils or linked arrays of receptors) in the absence of ligand. In effect, a receptor is 'immobile' if its average migration speed is much less than that of the ligand.

MAKING TRANSPORT VECTORIAL OR REVERSIBLE

A transport system comprising a succession of receptors A,B,C… as above is likely to be reversible if the receptors are mobile; a ligand detaching from C is as likely to bind again to receptor A as to E, all other things (such as availability and dissociation constants) being equal. If the receptors are immobile then access can potentially determine the direction of flux; a ligand detaching from C can attach to A if it remains linked to B but not if it is linked to D, but to E and not A if it is linked to D. Therefore, although the rate constants of individual ligand-receptor associations and dissociations are unlikely to have major effects on the overall transport rate (above), changes in them could well affect the net direction of transport (Fig. 7.3). In principle, transport could be made centripetal or centrifugal simply by modifying a subset of the receptors so that the kinetics of complex formation were altered appropriately. Some such mechanism could in principle underpin the nucleocytoplasmic shuttling of certain RNAs and proteins.

Almost certainly another contributing factor in vectorial transport is to make at least one step inherently vectorial, either by an energy-dependent net movement of ligand or by using a 'blocking factor' to prevent reassociation with a previous receptor. Some old and some more recent studies in mRNA translocation suggested that both these possibilities are realized when messengers (at least adenylated ones) pass the pore-complex.[29] The nucleoside triphosphatase seems to be vectorial because it is activated only by poly(A)$^+$ mRNA at the nuclear, not the cytoplasmic face,[30] and the major cytoplasmic poly(A) binding protein first described by Blobel[31] prevents reassociation of the messenger with P110.[24] It is not clear what this latter finding signifies in respect of the presumed association between P110 and its homologues and poly(A) within the cytoplasm. However, the consequences of this vectorial step are immediately evident; there are mass-action 'pull' and 'push' effects on transport events respectively preceding and succeeding it, and all other considerations aside, these will tend to make transport vectorial throughout the system. Of course, there is a danger that the final receptor, the terminal anchor for the ligand, will exert a contrary mass-action effect as its occupancy increases. Presumably

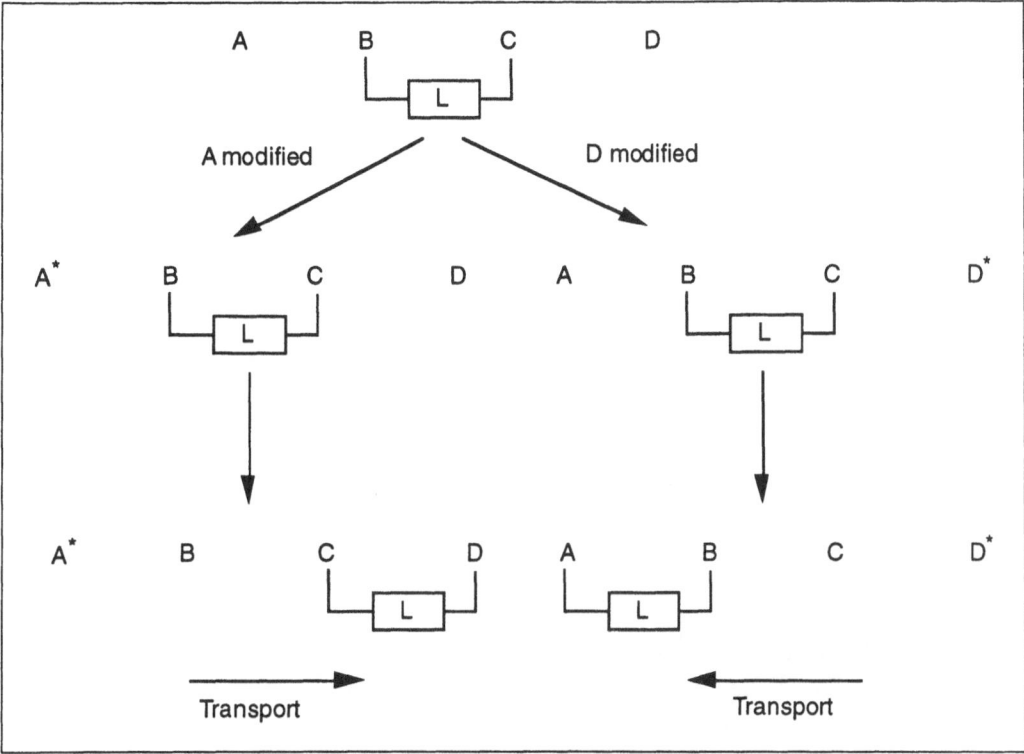

Fig. 7.3. Making transport vectorial by receptor modification. A sequence of receptors A-D carries ligand L, which simultaneously binds to two adjacent receptors. By modifying A or D transport can be directed forwards or backwards.

this is offset in practice by: (a) making the terminal receptor especially abundant and (b) blocking reassociation with the penultimate receptor by means of a suitably high-affinity ligand binding factor which comes into play after the ligand has been anchored.

Much of this mechanistic detail is suggested by commonsense considerations but is likely to remain speculative for some time to come; kinetic studies on receptor-ligand associations have been done for P110 and poly(A)[32] but are unlikely to be extended to other components of the system in advance of molecular biological and morphological studies. One final and highly speculative thought for the present on vectorial transport mechanisms concerns the potential role of cell water fluxes in reorientation. Generally, a ligand—especially one with a high axial ratio—is likely to reorientate itself in the direction of flow, a fact familiar to anyone who has ever dropped sticks into a river. It is feasible that protoplasmic streaming could be used to increase the efficiency of vectorial macromolecule transport when the transportants are not compactly folded.

GENERALIZING THE MODEL

Early chapters in this book were devoted to undermining some dubious presumptions about intracellular movement, especially in regard to macromolecules, and to exploring what the word 'transport' actually denotes. We have now arrived at a tentative model for nucleocytoplasmic transport of proteins and RNAs that conforms to a wide body of experimental data and is consistent with what we know about the character of the intracellular milieu. It obviously lacks detail, but it is capable of being elaborated through a sequence of testable hypotheses. Meanwhile, it puts the role of the pore-complex in proportion, making this structure important in the process but not exclusively so, and indicates how events at the pore-complex might be integrated with those within the nucleoplasm and cytoplasm. Even reviews that are predicated on pore-complex-only presumptions are now beginning to shift to this kind of question.[33] We should now inquire whether the model is more generally applicable.

In one respect, the answer is probably 'yes'. There is no a priori reason to suppose that transcytoplasmic protein migration involves fundamentally different mechanisms if the movement is directed towards nuclear import; importin and its homologs are no doubt key determinants of nuclear uptake, but probably the receptor-ligand-receptor pattern that we have proposed applies more generally, and intracytoplasmic movement simply involves factors other than the importin family. Since the model implies that the 'leapfrogging' of ligand between receptors is a major determinant of transport rate, it is likely that anything that affects this process will change experimental measurements of protein mobility in the cytoplasm. Heating will accelerate random ligand movements and given the low affinities of the receptors it will probably tend to decrease the chances of complex formation and therefore to slow down protein migration. This may explain why the so-called 'diffusivities' of cytoplasmic proteins, far from increasing linearly with temperature as the Einstein-Smoluckowski mechanism of diffusion requires, tend to decline as the temperature rises,[34] a phenomenon so apparently incompatible with physical principles and so resistant to explanation that is has hitherto been largely ignored.

Affinity of proteins for binding sites presumably has a negative temperature coefficient (log $K_d \propto {}^1/_T$) while the rate of transfer between receptors should have a positive one (log rate $\propto -{}^1/_T$). The net effect could be a virtual temperature dependence of migration. The agreement between the model and an otherwise intractable observation may be considered an argument in the model's favor. Dextrans show a more conventional temperature dependence of diffusivity,[34] presumably because they do not engage in transient receptor crosslinking. However, their migration should not be interpreted simplistically; the cytoplasm through which they pass still has gel-like properties, and 'diffusion' through gels is notoriously difficult to model even when it is permissible to assume that Brownian motion is a major contributor to movement.[35]

Similar considerations apply to the transmission of information from environmental signals at the cell surface to the internum, including the nucleus. Recently Bray[36] has suggested that information transduction and communication systems comprising receptors, G-proteins, kinases, phosphatases and targets should be conceived as neural networks in which each unit is a protein capable of acting as a logic gate by virtue of its ability to switch between different activity states. It is evident that although some intracellular protein circuits of this kind might be permanent (as when the cytoskeleton acts as an information channel), most of them are ephemeral because the units are mobile. We demur at Bray's assumption that diffusion accounts for their movements over more or less long distances,[36] but we accept that over a short range (tens of nanometers, by the arguments in chapters 1 and 2) Brownian motion operates. If it did not, then it is difficult to see how any association-dissociation processes could occur. In any case, thermal 'noise' in the exact location of a circuit unit is likely to increase the pattern storage capacity of the network by stochastic resonance, so long as that noise is not excessive.

Our detailed critique of diffusion theory applied to biology[37] ended with a provisional distinction between two types of intracellular transport, solid-state and hydrodynamic. We proposed that the former applied to mRNA and perhaps other macromolecules, while the latter, which depended on protoplasmic streaming, applied to low M_r or inert solutes.[38] We must now consider a less clear-cut distinction. The former perspective, so we have reasoned in this book, needs to be replaced by a model in which solid-phase networks are transient and depend for their ephemeral integrity on the presence of the transportant itself. It is at least possible that fluid-phase fluxes contribute to vectorial migration in a system described by this model. The later perspective needs to take into account the presence of solid-phase structures; not only the exclusion zones represented by cytoskeleton, cytomembranes and perhaps the vicinal water surrounding those structures, but also the misty barriers formed by the transient associations of migrating macromolecules. The style of thinking implicit in the model we have proposed entails a convergence of perspectives that previously seemed distinct. Without doubt the movement of low M_r solutes between nucleus and cytoplasm must, if this inference is accepted, be affected by concurrent macromolecular traffic.

We can picture a solute molecule being swept along cytoplasmic channels by bulk solvent flow through ever-shifting gel barriers, passing through pore-complexes that are in the open state in consequence of protein or RNA ingress or egress, and entering the canals of the extrachromosomal network. Simultaneously, the macromolecules ascend and descend arrays of largely immobilized receptors, aided or hindered by fluid currents, much as climbers seeking successions of handholds and footholds on a rock face are affected by airflow; and the handholds and footholds themselves change their availabilities and accessibilities (receptors can be occupied or modulated), directing net movement now this way, now that, targeting the climbers to various destinations where they can be anchored.

In the light of the current balance of evidence, we believe that it is in a picture of this kind that "the meaning of nucleocytoplasmic transport" lies; or, to be bolder, that "the meaning of intracellular transport" lies.

References

1. Agutter PS. Models for solid-state transport: messenger RNA movement from nucleus to cytoplasm. Cell Biol Internat 1994; 18:849-858.
2. Jackson DA, Cook PR. Transcription occurs at a nuclear cage. EMBO J 1985; 4:919-926.
3. Zeitlin S, Parent A, Silverstein S et al. Pre-mRNA splicing and the nuclear matrix. Mol Cell Biol 1987; 7:111-120.
4. Feldherr CM. Ribosomal RNA synthesis and transport following disruption of the nuclear envelope. Cell Tiss Res 1980; 205: 157-162.

5. Lawrence JB, Singer RH, Marselle LM. Highly localized tracks of specific transcripts within interphase nuclei visualized by in situ hybridization. Cell 1989; 57:493-502.

6. Huang S, Spector DC. Nascent pre-mRNA distribution within the chromatin-depleted nuclear substructure demonstrated by in situ hybridization coupled with biochemical fractionation. Genes Dev 1991; 5:2288-2302.

7. Xing Y, Lawrence J. Preservation of specific RNA distribution within the chromatin-depleted nuclear substructure demonstrated by in situ hybridization coupled with biochemical fractionation. J Cell Biol 1991; 112:1055-1063.

8. Long BH, Huang C-Y, Pogo AO. Isolation and characterization of the nuclear matrix in Friend erythroleukemia cells: chromatin and the HnRNA associations with the nuclear matrix. Cell 1979; 18:1079-1090.

9. Agutter PS, Birchall K. Functional differences between mammalian nuclear matrix and pore-lamina preparations. Exp Cell Res 1979; 124:453-460.

10. Berezney R. Fractionation of the nuclear matrix I. Partial separation into matrix protein fibrils and a residual ribonucleoprotein fraction. J Cell Biol 1980; 85:641-650.

11. Dreyer C, Stick R, Hausen P. Uptake of oocyte proteins by nuclei of *Xenopus* embryos. In: Peters R, Trendelenberg M, eds. Nucleo-Cytoplasmic Transport. Heidelberg, Springer-Verlag, 1986:143-157.

12. Zachar Z, Kramer J, Mims IP et al. Evidence for channelled diffusion of pre-mRNAs during nuclear RNA transport in metazoans. J Cell Biol 1993; 121:729-742.

13. Van Venrooij WJ, Sillekens PTG, van Eekelen CAG et al. On the association of mRNA with the cytoskeleton in uninfected and adenovirus infected human KB cells. Exp Cell Res 1981; 135:79-92.

14. Davis KI, Ish-Horowicz D. Apical localization of pair-rule transcripts requires 3' sequences and limits protein diffusion in *Drosophila* blastoderm embryo. Cell 1989; 67:927-940.

15. Singer RH. The cytoskeleton and mRNA localization. Curr Opin Cell Biol 1992; 4:15-19.

16. Otegui C, Patterson RJ. RNA metabolism in isolated nuclei: processing and transport of immunoglobulin light chain sequences. Nuc Acids Res 1981; 9:4676-4681.

17. Agutter PS. An assessment of some methodological criticisms of RNA efflux studies using isolated nuclei. Biochem J 1983; 214:915-921.

18. Feldherr CM, Ogburn JA. Mechanisms for the selection of nuclear polypeptides in *Xenopus* oocytes. II: Two-dimensional gel analysis. J Cell Biol 1980; 87:589-593.

19. Paine PL. Diffusive and non-diffusive proteins in vivo. J Cell Biol 1984; 99:188s-195s.

20. Meier UT, Blobel G. Nopp140 shuttles on tracks between nucleo-lus and cytoplasm. Cell 1992; 70:127-138.
21. Paine PL. Nuclear protein accumulation by facilitated transport and intranuclear binding. Trends Cell Biol 1993; 3:325-329.
22. Srer P. The metabolon. Trends Biochem Sci 1985; 10:109-113.
23. Srivastava DK, Bernhard S. Metabolite transfer via enzyme-enzyme complexes. Science 1986; 234:1081-1086.
24. Bernd A, Schröder HC, Zahn RK et al. Modulation of nuclear envelope NTPase by poly (A) rich mRNA and by microtubule pro-teins. Eur J Biochem 1982; 129:43-49.
25. Dworetzky SI, Lanford RE, Feldherr CM. The effects of variations in the number and sequence of targeting signals on nuclear uptake. J Cell Biol 1988; 107:1279-1288.
26. Fey EG, Wan KM, Penman S. Epithelial cytoskeletal framework and nuclear matrix/intermediate filament scaffold: three-dimensional organization and protein composition. J Cell Biol 1984; 98: 1973-1984.
27. Georgatos SD. Towards an understanding of nuclear morphogen-esis. J Cell Biochem 1994; 55:69-76.
28. Van Eekelen CAG, van Venrooij WJ. HnRNA and its attachment to a nuclear matrix. J Cell Biol 1981; 88:554-563.
29. Prochnow D, Thomson M, Schröder HC et al. Efflux of RNA from resealed nuclear envelope ghosts. Arch Biochem Biophys 1994; 312:579-587.
30. Prochnow D, Riedel N, Agutter PS et al. Poly (A) binding pro-teins located at the inner surface of resealed nuclear envelope vesicles. J Biol Chem 1990; 256:6536-6541.
31. Blobel G. A protein of molecular weight 73000 bound to the polyadenylate regions of eukaryotic messenger RNA. Proc Natl Acad Sci USA 1973; 924-928.
32. Schröder HC, Diehl-Seifert B, Rottmann M et al. Functional dis-section of the nuclear envelope mRNA translocation system: effects of phorbol ester and a monoclonal antibody recognizing cytoskeletal structures. Arch Biochem Biophys 1988; 261:394-404.
33. Fabre E, Hurt EC. Nuclear transport. Curr Opin Cell Biol 1994; 6:335-342.
34. Peters R. Fluorescence microphotolysis to measure nucleocytoplasmic transport and intracellular mobility. Biochim Biophys Acta 1986; 864:305-359.
35. Crank J. The Mathematics of Diffusion. Oxford: Clarendon, 1975.
36. Bray D. Protein molecules as computational elements in living cells. Nature 1995; 376:307-312.
37. Agutter PS, Malone PC, Wheatley DN. Intracellular transport mechanisms: a critique of diffusion theory. J Theoret Biol 1995; 176:261-272.
38. Remenyik CJ, Kellermeyer M. A fluid mechanical hypothesis for

macromolecular transport in living cells. Physiol Chem Phys 1978; 10:107-113.

INDEX